PROTECTING WHAT MATTERS

PROTECTING WHAT MATTERS

Technology, Security, and Liberty since 9/11

CLAYTON NORTHOUSE

editor

COMPUTER ETHICS INSTITUTE

BROOKINGS INSTITUTION PRESS
Washington, D.C.

Protecting What Matters: Technology, Security, and Liberty since September 11 may be ordered from Brookings Institution Press, c/o HFS, P.O. Box 50370, Baltimore, MD 21211-4370; Tel.: 800/537-5487, 410/516-6956; Fax: 410/516-6998; Internet: www.brookings.edu

Library of Congress Cataloging-in-Publication data
    Protecting what matters : technology, security, and liberty since
      September 11 / Clayton Northouse, editor.
        p.   cm.
    Includes bibliographical references and index.
    ISBN-13: 978-0-8157-6126-6 (cloth ed. : alk. paper)
    ISBN-10: 0-8157-6126-0 (cloth ed. : alk. paper)
    ISBN-13: 978-0-8157-6125-9 (paper ed. : alk. paper)
    ISBN-10: 0-8157-6125-2 (paper ed. : alk. paper)
      1. Liberty—United States. 2. Civil rights—United States. 3. National
    security—United States. 4. Electronic surveillance—United States.
    5. Information technology—United States. 6. Terrorism—United
    States—Prevention. 7. September 11 Terrorist Attacks, 2001—Influence.
    I. Northouse, Clayton. II. Computer Ethics Institute. III.Title.
    JC599.U5P76 2006
    323.44'80973—dc22   2006002105

9 8 7 6 5 4 3 2 1

The paper used in this publication meets minimum requirements of the American National Standard for Information Sciences—Permanence of Paper for Printed Library Materials: ANSI Z39.48-1992.

Typeset in Sabon

Composition by OSP, Inc.
Arlington, Virginia

Printed by R. R. Donnelley
Harrisonburg, Virginia

Contents

Part III. Technology, Security, and Liberty: The Legal Framework

Foreword

The balance between civil liberties and national security is surely the issue of our time, and computers and information technology stand foursquare at the heart of this debate. As computers have penetrated our everyday lives and the Internet has become the central highway of our shared knowledge, the ethical and policy issues generated by information technology have inevitably moved to center stage. The new information technologies can be a powerful force for positive change, but they also create new risks. The Internet was one of the main communications mechanisms for terrorists in the days and weeks leading to September 11; now it is a primary tool for those entrusted with tracking and capturing terrorists.

In exploring the dangers and opportunities created by such new technologies, this anthology aims to illuminate the tensions that arise when a society is forced to rethink its fundamental values in the face of growing challenges to its security. To help address these issues, this volume calls on a veritable who's who of technical, legal, and public policy gurus.

How did the Computer Ethics Institute become involved in this project? CEI's mission is to provide a forum and resource for identifying, assessing, and responding to ethical issues associated with the advancement of information technologies. Both locally and globally, computer and information technologies have become the infrastructure of business and government, transforming the relationships among government, businesses, stakeholders, and the communities within which these relationships take place. Because of these relationships, it is critically important to formulate organizational and public policies that create the optimal balance among corporate, public, and private welfare.

The Brookings Institution has been a partner of the Computer Ethics Institute since its inception, serving as the venue for CEI conferences, hosting the CEI website, and providing support whenever necessary. In fact, it was a CEI event—held at Brookings and cosponsored by Ascential Software, the Computer Ethics Institute, and the Brookings Institution—that gave rise to this book. The event, *Balancing Civil Liberties and National Security in the Post-9/11 Era: The Challenge of Information Sharing,* brought together technologists, government officials, legislators, and public policy scholars to focus on the following three questions: How can information technologies assist in maintaining a secure homeland? What issues—legal, cultural, ethical, and organizational—may arise from the implementation of these information technology solutions? And what operational framework should policymakers use to maximize the benefits and minimize the harm of implementing these solutions in the post–September 11 environment?

There were many different responses to these questions. We heard of the need to develop strong leadership with a willingness to define expectations and deadlines. We were cautioned to delineate specific goals, avoid generalities, and look at alternative frameworks, particularly those that do not pit the government against the people or the world. It was noted that we should carefully identify what we need to know and what good intelligence is. We were also urged to take privacy seriously and work toward consensus. Finally, we were warned not to depend solely on the judiciary to balance civil liberties and security.

Before arriving at answers, we were reminded that we need to find out what systems currently exist to deal with these issues. Then, to develop an appropriate framing of the questions and issues involved, we could bring together groups, including representatives not just from the federal government but also from state and local governments and the private sector. For best results, we were advised to operate more tactically than philosophically, to establish a research agenda, and to provide education.

All of these were important observations, but the one that truly stood out was the need to depolarize the issue so that all participants could build from common ground. It was abundantly clear from the beginning that this is a very controversial issue. In fact, the only way to bring these participants together was through an event that was by invitation only, not for attribution, and free of the press.

After that event and some follow-up activities, we clearly needed a publication, bringing together expert thought from the relevant fields, to illuminate the issue. That has been the monumental task of Clayton Northouse, the editor, who brings to this project the mind of the philosopher and the passion of the practitioner. He has worked with the people, the topic, and the book from the early days of his involvement with CEI and has now actually put it together. It has been a labor of love and a contribution to us all.

Now everything else is in the hands of the readers. We trust that our time and effort will be well received.

RAMON C. BARQUIN
President, Computer Ethics Institute
President, Barquin International

JANE FISHKIN
Vice President, Computer Ethics Institute
Vice President, Brookings Institution

Acknowledgments

With any anthology there are obviously a lot of people who contribute to the making of the final product, namely the contributors, but there are many people who go unmentioned. I first proposed the idea of doing a book on national security and civil liberties to Robert Faherty, the director of Brookings Institution Press. He liked the idea and encouraged its development. Jane Fishkin and Cynthia Darling also offered a great deal of support from within Brookings. This work would not have happened were it not also for the support of Ramon Barquin, who helped me in every step of the process. My parents, Cam and Donna Northouse, offered advice throughout the process. Mary Kwak, acquisitions editor at Brookings Institution Press, contributed enormous effort in the editing and shaping of the final product. And, lastly, Zoë Baird, James Barksdale, Bruce Berkowitz, Jerry Berman, Senator Russ Feingold, Beryl Howell, Senator Jon Kyl, Gilman Louie, James Steinberg, Larry Thompson, Gayle von Eckartsberg, and Alan Westin made the anthology what it is.

I

Introduction:
Security and Liberty in the
Twenty-first Century

1

Providing Security and Protecting Liberty

CLAYTON NORTHOUSE

On November 9, 2002, readers of the *New York Times* learned that Pentagon researchers planned to develop a massive virtual database, potentially containing data on every American, that could provide "instant access to information from Internet mail and calling records to credit card and banking transactions and travel documents."[1] Known as Total Information Awareness (TIA), the program originated in the Defense Department's Information Awareness Office, which was set up after September 11 to help develop predictive technologies that could aid the government in preventing future attacks. TIA planners hoped to exploit the vast amount of electronic information stored in commercial and governmental databases to find and track terrorists. Their goal was to develop analytical tools that would search through these mountains of data and generate an electronic profile of likely terrorists. Taken together, these tools and the databases to which they were applied could provide the government with an all-seeing eye on the world. In fact, TIA's logo was the all-seeing eye found on the U.S. dollar bill, and its motto was "scientia est potentia" (knowledge is power).

News of TIA unleashed a firestorm of protest, not only among left-leaning civil libertarians but also on the right. A few days after news of TIA broke, William Safire used his *Times* column to warn:

Every purchase you make with a credit card, every magazine sub-scription you buy and medical prescription you fill, every Web site you visit and e-mail you send or receive, every academic grade you receive, every bank deposit you make, every trip you book and every event you attend—all these transactions and communications will go into what the Defense Department describes as "a virtual, centralized grand database."

To this computerized dossier on your private life from commer-cial sources, add every piece of information the government has about you—passport application, driver's license and bridge toll records, judicial and divorce records, complaints from nosy neigh-bors to the F.B.I., your lifetime paper trail plus the hidden camera surveillance—and you have the supersnoop's dream: a "Total Infor-mation Awareness" about every American citizen.[2]

Other analysts came to TIA's defense, arguing that the new security challenges facing the United States demanded a new type of response. Since the end of the cold war, nonstate actors had replaced foreign gov-ernments as the major threats to U.S. national security. In order to track and defeat enemy combatants in decentralized networks spanning the globe, the intelligence community had to collect more data than ever before and draw links between seemingly innocuous bits of information. "It is the only way to protect ourselves," explained former CIA official John MacGaffin. "For the last forty years, there were a finite number of bad guys coming out of a finite number of places. Now we have an infi-nite number of threats from an infinite number of things."[3]

Tools such as Total Information Awareness, advocates maintained, were critical to this task. In addition, they argued, concerns about civil liberties could be addressed by ensuring that privacy safeguards were in place. A combination of judicial oversight and modern technologies, such as anonymity tools, could allow the government to fight the war on ter-ror without infringing unduly on ordinary citizens' rights.

The public and Congress were not convinced. For many, TIA's Big Brother overtones were too difficult to ignore. And the fact that the Infor-mation Awareness Office was headed by retired rear admiral John M.

Poindexter, who was convicted of lying to Congress about the Iran-Contra affair, did little to allay their concerns. In May 2003 the Defense Department responded to TIA's critics by releasing a detailed report, as required by Congress, that pledged to make the protection of Americans' privacy and civil liberties a "central element" of the program. It also announced that henceforth TIA would be known as Terrorist Information Awareness rather than Total Information Awareness. But the critics were unappeased. Even Poindexter's resignation later that summer—following a new flap over plans to launch a terrorism futures trading market—failed to quell opposition to the programs he had helped create. In September 2003 Congress cut off funding for TIA and shut down the Information Awareness Office.

This case illustrates the controversy provoked by ambitious efforts to harness information technology to the cause of homeland defense. It also raises a number of questions that will remain vital long after this particular program's demise: What principles should guide us in negotiating the relationship between security and liberty in the aftermath of September 11? How does technology factor into this complex set of concerns? What benefits do techniques like data mining offer, and how should they be used? To what extent are we willing to give the government control over our personal information? Do current efforts to exploit information and information technology violate the principles embodied in the Fourth Amendment?

The contributors to this volume address these critical questions. In the next essay in this section, Alan Westin examines public opinion data to identify the broad contours of the current debate over security, liberty, and technology. In the second section, "Protecting Security and Liberty: Information Technology's Role," James Steinberg, Zoë Baird, James Barksdale, Gilman Louie, Gayle von Eckartsberg, and Bruce Berkowitz analyze the necessary restructuring of the intelligence community and the role that technology can play in combating terrorism. They also suggest how technology can be used to protect the homeland without necessarily threatening civil liberties. Finally, in the third section, "Technology, Security, and Liberty: The Legal Framework," Larry Thompson, Jerry Berman, Beryl Howell, Senator Jon Kyl, and Senator Russell Feingold focus on key legal issues at the intersection of liberty and security and continue the debate over the proper legal restrictions on the government's power to use information technologies for national security purposes.

Security and Liberty: The Fundamental Debate

American history is, to a great extent, a study of the tension between liberty and security. The Founders' desire to protect what they saw as inalienable rights, including liberty of thought, association, and speech and freedom from unwarranted government incursions into citizens' homes, is enshrined in the Bill of Rights. Yet over the more than two hundred years since the Constitution was ratified, these basic liberties have been compromised repeatedly during periods of national uncertainty.[4]

In 1798, with the nation prepared for war, President John Adams and the Federalists passed the Alien and Sedition Acts, which made any "false, scandalous, and malicious" statement against the United States government punishable by fine and imprisonment. In addition, they gave the president the exclusive authority to deport any foreigner considered to be a threat to national security. At a time of rising tensions with France, the Federalists argued that these measures were necessary to preserve order and protect the nation. But in practice, they were used primarily to muzzle the opposition Republican Party. Nearly all of the newspaper writers and editors arrested under the Alien and Sedition Acts were Republicans. The acts expired on the last day of Adams's presidency, and his successor, Thomas Jefferson, released and pardoned all those jailed as a result of this legislation. The Alien and Sedition Acts have since become a black mark in the history of free speech in America and the subject of condemnation by the Supreme Court.

Some sixty years later, in the midst of the Civil War, President Lincoln faced opposition to Union forces in Baltimore. When rioting broke out among Confederate sympathizers, resulting in the death of several Union soldiers, Lincoln suspended the writ of habeas corpus, which gives detained individuals the right to have their case heard before a judge, and declared the entire state of Maryland under martial law. Throughout the war, Lincoln suspended the writ of habeas corpus eight times, finally issuing a nationwide order. As a result, thousands of supposed Southern sympathizers, draft dodgers, and deserters were detained without access to a civilian court of law. After the end of the Civil War, the Supreme Court condemned these actions, ruling in *Ex Parte Milligan* that it was unconstitutional to detain a U.S. citizen under martial law without access to functioning civilian courts.[5]

The next major challenge to Americans' civil liberties came during World War I. With the Espionage Act of 1917 and the Sedition Act of

1918, the United States returned to many of the practices authorized under the Alien and Sedition Acts. In the first of what would become two Red Scares, thousands of individuals were arrested for speaking out against the war and for criticizing the United States government. At the time the Supreme Court upheld a number of decisions involving the detention of individuals who had opposed the war, but all of these decisions were subsequently overturned, and every individual arrested under the Espionage and Sedition Acts was eventually released.

Later, during World War II, widespread panic up and down the West Coast led to the internment of 120,000 people of Japanese descent. Under tremendous political pressure, President Roosevelt issued Executive Order 9066 ten weeks after the attack on Pearl Harbor. This order gave the Army the power to establish military zones from which certain individuals could be excluded. Ninety percent of Japanese Americans were uprooted from their communities, forced to leave their homes and businesses, and relocated to internment camps in which they remained for up to three years. In *Korematsu* v. *United States,* which was decided in 1944, the Supreme Court upheld this policy. Writing for the majority, Justice Hugo Black stated that "the power to protect must be commensurate with the threatened danger."[6] Since then, several presidents have apologized for the forced internment of the Japanese, and the Supreme Court has never relied on *Korematsu* as a precedent in deciding later cases.

Perhaps most famously, at the height of the cold war, the Red Scare of the 1950s involved the blacklisting of hundreds of supposed Communist sympathizers and the incarceration of Communist Party leaders. The House Un-American Activities Committee blacklisted hundreds of artists and writers, and under the Smith Act, members of the Communist Party were prosecuted for conspiring to overthrow the U.S. government. In *Dennis* v. *United States,* the Supreme Court affirmed the constitutionality of the Smith Act and upheld the conviction of Eugene Dennis and ten other Communist Party leaders, declaring their speech to pose a clear and present danger. In a strong dissenting opinion, Justice Black observed, "Public opinion being what it now is, few will protest the conviction of these Communist petitioners. There is hope, however, that in calmer times, when present passions and fears subside, this or some later Court will restore the First Amendment liberties to the high preferred place where they belong in a free society."[7] Eventually, the Court vindicated Black's hopes and brought the second Red Scare to an end by restricting the scope of the Smith Act and prohibiting Congress from investigating people's political beliefs.

As these examples demonstrate, the issues raised by the sometimes conflicting demands of security and liberty are not new. But today the controversy surrounding the relationship between these two goals is heightened by the advent of powerful and, to some, frightening new technologies. Cameras can now record the geometric structure of a subject's face and instantly compare those measurements against data on suspected criminals. Giant databases can store information on every credit card transaction, medical record, bank account, and plane reservation. Analysts can perform clandestine searches of the data stored on individuals' computers and collect data transmitted over the Internet, including the addresses to which e-mail is sent and the websites that a user has visited. In some eyes these capabilities evoke the Orwellian nightmare of a paternalistic, omnipotent government that observes its citizens' every move.

How will the government respond to the civil liberties challenges that these new technologies raise? In large part the answer to this question lies in public beliefs about how the balance between security and liberty should be struck. As Learned Hand said, "Liberty lies in the hearts of men and women; when it dies there, no constitution, no law, no court can save it; no constitution, no law, no court can even do much to help it."[8] Accordingly, in the following essay, Alan Westin offers a detailed examination of the public's attitudes toward civil liberties and national security before and after September 11. Based on the results of five surveys conducted since the September 11 attacks, he finds that large majorities both support the government's expanded powers and remain concerned about safeguarding civil liberties. This attitude of "rational ambivalence," he concludes, should be seen as an opportunity to ensure that both support for antiterrorist programs and protections for civil liberties remain strong.

Protecting Security and Liberty: Information Technology's Role

The second section of *Protecting What Matters* focuses on the intelligence challenges posed by terrorism and the role that information technology can play in this new threat environment. During the cold war, as James Steinberg points out, the task facing the intelligence community was relatively straightforward: "We generally knew what to look for and where to look for it." Moreover, most of the necessary information concerned military activities overseas, and the expertise needed to collect and analyze it resided in the federal government. Since September 11 all that has changed. In his essay Steinberg discusses how the intelligence community must adapt to meet future security challenges. He also identifies new technologies that can aid in this task, as well as tools that can promote accountability in the collection and use of sensitive personal information.

Zoë Baird and James Barksdale, cochairs of the Markle Task Force on National Security, focus on one of the most important aspects of the new intelligence challenge: the need to improve information sharing across different agencies and levels of government. Based on the task force's work, they outline six criteria that an effective Systemwide Homeland Analysis and Resource Exchange (SHARE) network must meet and analyze the proposed network's technological components. They also review recent policy developments—notably the executive orders issued by President Bush in August 2004 and the Intelligence Reform and Terrorism Prevention Act of 2004—that create a national framework for better information sharing and, ultimately, greater security.

Gilman Louie and Gayle von Eckartsberg also explore information technology's role in the post–September 11 world, but their emphasis is on tools that make it possible to protect civil liberties and the nation at the same time. For example, selective revelation and anonymizing technologies can limit violations of privacy while granting the government access to a great deal of useful information. The availability of such techniques, Louie and von Eckartsberg argue, makes security-versus-liberty a false choice.

Finally, Bruce Berkowitz looks beyond specific technologies to delineate the role of policies and procedures in creating a safe zone for collecting and sharing information while protecting civil liberties. "Since September 11," he writes, "the main approach to resolving these problems has been to 'lower the bar'—that is, reduce the barriers that preclude intelligence and law enforcement agencies from investigating individuals and sharing information." Instead, he argues, the government should "adopt measures that limit the potential damage of such investigations." By limiting the mandate of information collectors, controlling the use of information, and providing recourse for the subjects of mistaken investigations, the intelligence community can more effectively take advantage of technology's potential without unduly infringing upon individual rights.

Technology, Security, and Liberty: The Legal Framework

The essays in the final section of *Protecting What Matters* analyze the legal context for the current debate on technology, security, and liberty. Four overlapping areas of law are relevant to this discussion. First, the Fourth Amendment provides protection against unreasonable searches and seizures, whether physical or electronic. Title III of the Omnibus Crime Control and Safe Streets Act of 1968 builds on this foundation by setting forth the procedures the government must follow to obtain a warrant for electronic surveillance. Second, a weaker set of regulations controls governmental access to information voluntarily conveyed to third parties, such as checking account records or the telephone numbers of incoming and outgoing calls. Third, the Foreign Intelligence Surveillance Act of 1978 controls the government's use of electronic surveillance to collect foreign intelligence for national security purposes. Finally, a disjointed collection of privacy legislation governs the use of personal information in the public and private sectors. The remainder of this introduction provides a brief overview of these four areas of law as background for the more detailed analyses presented by the volume's contributors. It also discusses the changes introduced by the Uniting and Strengthening America by Providing Appropriate Tools Required to Intercept and Obstruct Terrorism Act of 2001 (USA PATRIOT Act).

The Fourth Amendment and Title III

Much of the law guiding the debate over civil liberties and national security centers on the Fourth Amendment, which protects the "right of the people to be secure in their persons, houses, papers, and effects, against unreasonable searches and seizures." In *Katz* v. *United States*, the Supreme Court extended this right to include protection against electronic intrusions.[9] As Justice Potter Stewart wrote for the majority,

> The Fourth Amendment protects people, not places. What a person knowingly expresses to the public, even in his own home or office, is not a subject of the Fourth Amendment protection. . . . But what he seeks to preserve as private, even in an area accessible to the public, may be constitutionally protected . . . once it is recognized that the Fourth Amendment protects people—and not simply "areas"— against unreasonable searches and seizures, it becomes clear that

the reach of that Amendment cannot turn upon the presence or absence of a physical intrusion into any given enclosure.[10]

Justice Stewart also observed that "searches conducted outside the judicial process, without prior approval by judge or magistrate, are per se unreasonable under the Fourth Amendment—subject only to a few specifically established and well-delineated exceptions."[11] Consequently, in order to meet the constitutional test, the procedures for authorizing electronic search warrants must be clearly established, and government wiretapping must receive prior approval from a judge.

In response to *Katz*, Congress enacted the Omnibus Crime Control and Safe Streets Act of 1968 (OCCSSA). Title III of OCCSSA prohibits warrantless wiretapping of electronic, telephone, and face-to-face conversations and establishes procedures regulating the use of wiretaps. For a limited set of criminal offenses, including murder, kidnapping, extortion, gambling, and drug sales, a judge or magistrate can authorize the Department of Justice (DOJ) to eavesdrop on conversations for up to thirty days. After this period the courts are bound to notify those whose conversations were monitored. To obtain authorization for a wiretap under Title III, the DOJ must, among other things, prove that there is probable cause to believe that the targeted person committed or is about to commit one of the criminal offenses.

In the decades since the enactment of OCCSSA, the government has faced the constant challenge of keeping up with the advance of technology. In 1994 Congress passed the Communications Assistance for Law Enforcement Act (CALEA) to ensure that new communications technologies would permit eavesdropping by law enforcement agencies. However, CALEA's reality has never lived up to its promise. In his essay Larry Thompson argues that this piece of legislation needs to be more vigorously enforced. Otherwise, he warns, "The government may simply not have the technological ability or the capacity to undertake timely electronic surveillance"—making concerns about potential tradeoffs between security and liberty moot.

Access to Third-Party Information

The Fourth Amendment and Title III do not apply to documents and information voluntarily conveyed to third parties. The Supreme Court established this principle in *United States* v. *Miller*, which dealt with access to checks and other financial records held by a third party, such as a bank. The Court reasoned that there was no reasonable expectation of privacy in such a situation because the documents were "voluntarily conveyed to the banks and exposed to their employees in the ordinary course of business."[12] Consequently, no matter how sensitive the data—be it medical records, educational records, financial records—the government is allowed to search and seize documents voluntarily given to third parties without fear of violating the Fourth Amendment (although privacy legislation may bar the government from doing so in specific cases). To obtain a court order authorizing access to such records, the government need only show reasonable grounds for believing that the targeted information is relevant and material to a criminal investigation—a far lower hurdle than the one established by Title III.

The standard of evidence is even lower when it comes to "contentless" data routinely held by third parties, such as the information collected by pen registers and trap-and-trace devices. These technologies are like caller IDs, recording the telephone numbers for incoming or outgoing calls on a given line. In *United States* v. *New York Telephone Company* (1977), the Supreme Court found that Title III does not cover the use of pen registers and trap-and-trace devices because the content of the conversations is not captured, only the phone numbers being used.[13] Two years later, in *Smith* v. *Maryland*, the Court ruled that the Fourth Amendment offers no protection against the government's use of these devices.[14] The Court argued that there was no reasonable expectation of privacy in such cases because it is common knowledge that telephone companies record these numbers in the normal course of business. (For example, the numbers dialed are printed on phone bills.) Consequently, the government can receive a court order for the use of these devices by simply making a sworn declaration that the sought-after information is relevant to a criminal investigation. The orders and the information received never have to be revealed to their targets.

Foreign Intelligence

An entirely different legal regime—the Foreign Intelligence Surveillance Act of 1978 (FISA)—governs domestic surveillance of foreign powers or agents of foreign powers. FISA created two secret courts: the Foreign Intelligence Surveillance Court and the Foreign Intelligence Surveillance Court of Review. In order to receive a secret warrant for the collection of foreign intelligence information, the Department of Justice must demonstrate to the Foreign Intelligence Surveillance Court that there is "probable cause to believe that . . . the target of the electronic surveillance is a foreign power or an agent of a foreign power."[15]

Under FISA the bar for receiving a search warrant is set much lower than under Title III. For this reason, the legislation was carefully crafted to prevent prosecutors from using FISA to get around the more stringent requirements that apply to information gathering in ordinary criminal investigations.

Privacy Law

The final piece of the legal puzzle is an array of efforts to safeguard the privacy of personal information. The most prominent statute protecting privacy in the public sector is the Privacy Act of 1974, Congress's first attempt to control the federal government's collection, dissemination, and use of personal information. It applies to the use by all federal government agencies of "systems of records" or, in other words, any collection of records retrievable by an individual identifier, such as a name or Social Security number, that is under an agency's control.

The Privacy Act is based on four principles. First, federal agencies must give any American citizen or permanent resident access to any information stored about him or her. Second, agencies must follow "fair information practices" in handling and storing personal information. Among other things this means that the information collected must be accurate and necessary for an agency to fulfill its functions. Third, strict limits are set on an agency's ability to release individually identifiable information to other organizations or individuals. Finally, individuals can sue government agencies that fail to abide by these principles.

However, a number of limitations make the Privacy Act's protection of personal information less than complete. The term *agency* has been interpreted broadly in order to allow divisions within federal departments to share information. The Privacy Act also allows agencies to release data for "routine use," defined as a use that is compatible with the purposes for which the data were collected. The sharing of information for law enforcement purposes is exempt from the requirements of the act. And the Privacy Act does not apply to the government's use of data stored in the private sector, where only piecemeal legislation to protect privacy exists.

In his essay Jerry Berman highlights the lack of comprehensive legislation concerning the government's use of private sector data as a leading source of uncertainty for the public, the government, and the private sector. He also calls for the development of a new legal framework building on existing constitutional doctrine and fair information practices and argues that such rules "will not only protect civil liberties but will also enhance the effectiveness of government counterterrorism activities."

The PATRIOT Act

The PATRIOT Act of 2001 modified many of these areas of law. For example, sections 201 and 202 add terrorism, production or dissemination of chemical weapons, and computer crimes to the list of offenses that can be cited in requests to authorize wiretaps under Title III. Section 213 also gives the government the power to conduct a Title III search and seizure without contemporaneous notification of the target. If it is reasonable to think that the notification will have "an adverse effect," notice need not be given for a "reasonable" period of time. Such "sneak and peek" searches are not new. What makes section 213 controversial, though, is that "reasonable" is not defined. Moreover, under this provision, the government can seize, as well as search, property and communications without giving notice.

The PATRIOT Act also expanded government access to information held by third parties. Section 216 extends the regulations governing pen registers and trap-and-trace devices to allow the government to capture e-mail address and header information without notifying the target and without abiding by the strict principles of Title III. It also grants the government access to the URLs of the websites an individual has viewed. In addition, sections 216 and 220 permit nationwide orders for the interception of electronic communications. Previously, courts could only issue orders for the jurisdictions where they were located.

Perhaps most important, the PATRIOT Act loosened some of the legal restrictions that were designed to keep intelligence collection and domestic law enforcement distinct. Before the PATRIOT Act's passage, strict procedures governed the sharing of intelligence information with those responsible for criminal prosecutions. The criminal branch of the Department of Justice was prohibited from making recommendations for investigation and from directing or controlling intelligence-gathering activities. This separation of functions was reflected in FISA's requirement that "the purpose" of surveillance must be to capture foreign intelligence. However, the PATRIOT Act changed this wording to "a significant purpose." This change in language has permitted greater coordination between the criminal and intelligence branches of the DOJ. Critics warn that this shift has opened the possibility for abuse of FISA warrants since the Department of Justice can now monitor U.S. citizens who are believed to be agents of foreign powers, even if criminal prosecution is the investigation's primary goal.[16] In her contribution to this volume, Beryl Howell addresses these issues by reviewing FISA's legislative history, including the amendments introduced by the PATRIOT Act, with a focus on the change in FISA's "purpose" restriction. She also examines some of the problems that may arise as a result of this change and proposes steps both to strengthen FISA and to enhance public confidence in the law.

Finally, the PATRIOT Act has extended the government's intelligence-gathering powers in several important ways. For example, section 206 permits roving wiretaps on the target of a FISA search. The government can monitor all communications coming to or from the target without specifying the particular technologies that will come under scrutiny. This means the government may monitor public means of communication, such as public phones and computer terminals in libraries, causing many people who are not associated with the target to come under surveillance.

Spurring controversy among librarians and booksellers, section 215 allows the government to issue orders to obtain business records by certifying before the Foreign Intelligence Surveillance Court that such records are relevant to a terrorism investigation or clandestine intelligence activity. Furthermore, without court approval or congressional oversight, section 505 gives the FBI the power to issue National Security Letters (NSLs). NSLs are used to require that Internet service providers and telephone companies release web history, e-mail, and telephone information relating to a particular person relevant to a terrorism investigation. The targets of NSLs and section 215 orders are prohibited from disclosing to any third party the receipt of an order or the seizure of records.

In the two chapters that conclude this book, Senators Jon Kyl and Russ Feingold debate the value and significance of the PATRIOT Act. Senator Kyl argues that this legislation provides the necessary means for overcoming previous intelligence failures, without endangering civil liberties. Senator Feingold, the only member of the Senate to vote against the PATRIOT Act, takes an opposing view, maintaining that the powers it grants the government are overly broad.

Domestic Spying

In late 2005 the *New York Times* revealed that in the months following September 11, 2001, President Bush secretly authorized the National Security Agency (NSA) to spy on Americans without a warrant or court order.[17] Since then the NSA has been monitoring international phone calls and intercepting international e-mails between United States citizens and people in certain Middle Eastern countries.[18]

Two basic positions have been taken on the program's legality. The Department of Justice argues that President Bush acted at the "zenith of his powers in authorizing the NSA activities."[19] The American Civil Liberties Union, on the other hand, argues that the NSA program "seriously violates the First and Fourth Amendments" and is "contrary to the limits imposed by Congress."[20]

One of the central issues in this complex legal debate is whether the NSA program is in violation of the Foreign Intelligence Surveillance Act. As noted by the ACLU, when Congress enacted FISA, it also amended Title III of the Omnibus Crime Control and Safe Streets Act to state that the procedures of Title III and FISA "shall be the exclusive means by which electronic surveillance . . . and the interception of domestic wire, oral, and electronic communications may be conducted."[21] Hence, the ACLU concludes, because the NSA is acting outside of Title III and FISA procedures, it is in violation of the law.

The Department of Justice counters that section 109 of FISA "expressly contemplates that the Executive Branch may conduct electronic surveillance outside FISA's express procedures if and when a subsequent statute authorizes such surveillance."[22] The Authorization for Use of Military Force (AUMF) passed by Congress a week after September 11, 2001, authorized the president to "use all necessary and appropriate force" against those who attacked the United States. The Department of Justice argues that this gives the president the express authority to protect the nation and that a necessary component of protecting the nation is collecting intelligence on those who attacked the United States. Hence, Justice argues, the NSA program is consistent with FISA.

Former Senate majority leader Tom Daschle writes that he is "confident that the 98 senators who voted in favor of [AUMF] did not believe that they were also voting for warrantless domestic surveillance."[23] The Congressional Research Service (CRS) makes the additional point that, "Even if AUMF is read to provide the statutory authorization necessary to avoid criminal culpability under FISA, it does not necessarily follow that AUMF provides a substitute authority under FISA to satisfy the more specific language in Title III."[24] But CRS goes on to note that the legality of the NSA program is "impossible to determine without an understanding of the specific facts involved and the nature of the President's authorization, which are for the most part classified."[25]

These points will continue to be debated in Congress and before courts of law. In the process, the nation's laws and counterterrorism programs must adapt to the new environment created by the advancement of technology in the age of international terrorism. The chapters in this book sketch differing views on how these adjustments can take place as the government attempts to maximize the powers of information technologies to protect against terrorism while preserving civil liberties.[26]

Conclusion

How do we give the government the power to use technology for national security purposes while preserving our basic rights to privacy and freedom? New information technologies have the potential to be potent weapons in the war on terror. But if abused, they can also pose a significant threat to individual liberties. This challenge must be met head-on if the government is to succeed in its dual task of protecting liberty and providing security. The essays that follow do just that.

Notes

1. John Markoff, "Threats and Responses: Intelligence; Pentagon Plans a Computer System that Would Peek at Personal Data of Americans," *New York Times,* November 9, 2002, p. A12.

2. William Safire, "You Are a Suspect," *New York Times,* November 14, 2002, p. A35.

3. Siobhan Gorman, "Intelligence: Adm. Poindexter's Total Awareness," *National Journal,* May 8, 2004, p. 1430.

4. Geoffrey Stone, "Civil Liberties in Wartime," *Journal of Supreme Court History* 28, no. 3 (2003): 215–51; Geoffrey Stone, *Perilous Times: Free Speech in Wartime* (New York: W.W. Norton, 2004).

5. *Ex Parte Milligan,* 71 U.S. 2 (1866).

6. *Korematsu v. United States,* 323 U.S. 214 (1944).

7. *Dennis v. United States,* 341 U.S. 494, 580 (1951).

8. Learned Hand, *The Spirit of Liberty* (New York: Knopf, 1952), p. 190.

9. *Katz* v. *United States*, 389 U.S. 347 (1967). This decision rests in part on Justice Louis Brandeis's dissenting opinion in *Olmstead* v. *United States*, 277 U.S. 438 (1928).

10. *Katz* v. *United States*, 389 U.S 347, 351, 354 (1967).

11. Ibid., p. 357.

12. *United States* v. *Miller*, 425 U.S. 435 (1976).

13. *United States* v. *New York Telephone Company*, 434 U.S. 159 (1977).

14. *Smith* v. *Maryland*, 442 U.S. 735 (1979).

15. 50 U.S.C. 1801(e)(1).

16. As the Foreign Intelligence Surveillance Court of Review writes, "It can be argued . . . that by providing that an application is to be granted if the government has only a 'significant purpose' of gaining foreign intelligence information, the Patriot Act allows the government to have a primary objective of prosecuting an agent of a non-foreign intelligence crime." Ibid.

17. The story was first revealed by the *New York Times* in Jame Risen and Eric Lichtblau, "Bush Lets U.S. Spy on Callers without Courts," *New York Times*, December 16, 2005, p. 1A.

18. The *New York Times* reports that the NSA program has been collecting large amounts of information (Eric Lichtblau and James Risen, "Spy Agencies Mined Vast Data Trove, Officials Report," *New York Times*, December 24, 2005, p. 1A), whereas the Bush administration claims that the program is narrow in scope.

19. U.S. Department of Justice, "Legal Authorities Supporting the Activities of the National Security Agency Described by the President," January 19, 2006, p. 2 (see www.fas.org/irp/nsa/doj011906.pdf).

20. Complaint for Declaratory and Injunctive Relief, *American Civil Liberties Union, et al.* v. *National Security Agency*. U.S. District Court, Eastern District of Michigan Southern Division, January 17, 2006.

21. 18 U.S.C. section 2511(2)(f).

22. Ibid., p. 20.

23. Tom Daschle, "Power We Didn't Grant," *Washington Post,* December 23, 2005, p. A21.

24. Elizabeth B. Bazan and Jennifer K. Elsea, "Memorandum: Presidential Authority to Conduct Warrantless Electronic Surveillance to Gather Foreign Intelligence Information," Congressional Research Service, January 5, 2006, p. 43.

25. Ibid., pp. 42–43.

26. The chapters in this volume, written after the revelation of the NSA program, do not directly address the domestic spying issue, except for the chapter by Alan Westin.

2

How the Public Sees the Security-versus-Liberty Debate

ALAN F. WESTIN

Since September 2001 two polar positions on the tension between security and liberty have been competing for popular support:

—*The Security First Position*: If we do not modify some of our traditional constitutional norms limiting government powers, we will not be able to fight terrorism, function as a reasonably safe society, and enjoy our liberties.

—*The Liberty First Position*: If we reduce our liberties by granting the government sweeping and uncontrolled investigative and surveillance powers, we will weaken the constitutional system that has made our nation great.

While there are many intermediate and nuanced positions between these two poles, the thrust of the policy debate is conveyed by the positions of these two camps.

Where does the public stand in this debate? As of July 2005, almost fifty publicly released surveys of public opinion have probed aspects of this question. These surveys offer valuable readings of public attitudes on the balance that should be struck between security and liberty. Some of these surveys also offer valuable guidance to policymakers about how to proceed in the next phase of antiterrorist programs. These two subjects will be the focus of this article.[1]

What Shapes Public Views

Five primary factors generally shape public views on the security-versus-liberty debate:

—perceptions of the current terrorist threat and the likelihood of further attacks;

—perceptions of how well the government is dealing with the threats thus far and the methods being used;

—perceptions of how government antiterrorist programs are affecting valued civil liberties;

—underlying orientations toward general security and liberty issues; and

—basic orientations on political issues in general—which may be shaped by political philosophy, party identification, and demographic factors (such as race, gender, income, and education).

The public rarely gets its inputs on the first three factors from direct experience. In this way security-versus-liberty differs from some other public policy issues, such as racial discrimination, taxation, e-mail spam, and healthcare costs, where citizens feel the policy issues directly. When it comes to homeland security, the public relies primarily on two sources of facts and messages: mass media coverage and positions taken by interest groups to which people belong, such as professional, ethnic, and other organizations.

With these cautions in place, I begin by sketching the pre–September 11 structure of American public opinion on general liberty-versus-order tensions. Next, I report on five trendline surveys between 2001 and 2005 that I worked on with Harris Interactive. Finally, I draw conclusions about what I believe the public is coming to expect by way of security program reforms from the White House, Congress, and the courts.

Pre–September 11 Public Opinion Baseline

The American republic was founded with a cautionary, even hostile attitude toward government and a strong desire to protect citizens' liberties. The Constitution and Bill of Rights expressed these concerns by limiting the government's powers of investigation, surveillance, and prosecution, and by incorporating extensive due process rules into all government proceedings, both civil and criminal.

During the nation's first 200 years, the government's powers to combat crime and maintain internal security steadily expanded in response to new challenges posed by urbanization, crime, and national security. These expansions of government authority typically occurred during periods of national emergency (for example, the Civil War and World Wars I and II) and were upheld by the courts in periods of assumed internal security threats (such as the Palmer Raids in the 1920s and McCarthyism in the 1950s). However, many of these expanded powers were rolled back once a perceived emergency came to an end. Moreover, basic liberties—constitutional rights to privacy, equality, and due process—were also expanded, especially in the second half of the twentieth century, driven by court rulings, legislation, and public opinion. Against this backdrop, one can identify several trends in public opinion on security and liberty in the decades just before September 11.

Distrust of Government

Beginning in the 1960s, there was a sharp rise in general distrust of government and other social institutions or, to put it another way, a sharp drop in the general trust in government and social institutions that had characterized public opinion during the 1940s and 1950s. By April 2001 only 31 percent of the public said they thought the federal government could be "trusted to do what is right at least most of the time."[2]

Concern over Invasions of Privacy

Between the early 1970s and the mid-1990s, there was a sharp rise in public concern about threats to personal privacy. In 1995, 82 percent of Americans said they were "concerned about threats to personal privacy in America today," and 47 percent of these said they were "very concerned." When asked which type of privacy threat worried them most, 51 percent chose government and 43 percent business, with 6 percent not sure.[3]

Support for Law Enforcement Powers

At the same time, support for broad government investigative powers remained high. An April 2001 survey found that 56 percent approved of the FBI or law enforcement agencies intercepting telephone calls to and from people suspected of criminal activities, 55 percent approved of intercepting letters and packages sent to or from people suspected of crimes, and 54 percent approved of intercepting e-mails sent to or from suspected criminals.[4]

Ambivalence on Proposals for National IDs

Views of proposals for national ID systems in these years are particularly valuable in illustrating the nature of public attitudes on liberty issues. Surveys from 1990 to 2001 generally asked about national IDs in the context of specific applications, not in the abstract, since there was overwhelming public opposition to any general national ID number or card for U.S. citizens.

In 1990, for example, 56 percent of the public said that they opposed requiring all employees to have a national work identity card as a means of identifying illegal alien workers.[5] However, during the debate over creating a national health insurance system in 1994, 60 percent of a national sample said that they would favor a national healthcare identification number for each person. Public support for a health ID number rose to over 75 percent if safeguards were added, such as forbidding use of a health ID number for any purposes other than health insurance and giving individuals the right to sue for damages if such ID numbers were misused.[6]

Strong Desire for Safeguards

When asked in 2000 whether it was acceptable for law enforcement officials to obtain information about criminal suspects from commercial data sources, majorities of 86 to 92 percent said yes. At the same time, large majorities, ranging from 56 to 83 percent, felt that information privacy safeguards should be installed and followed if commercial services were to provide individual data to government or other commercial users.[7]

Harris-Westin Surveys, September 2001 to June 2005

With this pre–September 11 framework in mind, I turn to post–September 11 readings of public opinion on security and liberty issues. Harris Interactive conducted five trendline national telephone surveys on security-versus-liberty issues between September 2001 and June 2005. The surveys were conducted September 19–24, 2001; March 13–19, 2002; February 12–16, 2003; February 9–16, 2004; and June 7–12, 2005.[8] All of the surveys were done for the Harris Poll, a self-funded set of Harris surveys carried out since 1963 for wire service use and as the basis for syndicated columns authored by Harris's chairman or a senior Harris staffer. I served as academic advisor on these surveys and contributed a commentary to the columns written by either Harris Interactive chairman Humphrey Taylor or senior vice president of Harris Interactive Public Affairs and Government Research, David Krane.

Conceptual Framework

When we constructed the first of these surveys, we developed three sets of questions, which served as trend questions on each of the successive surveys. First, we listed ten "increased powers of investigation that law enforcement agencies might use when dealing with people suspected of terrorist activity, which would also affect our civil liberties." We asked respondents if they would favor or oppose each one. (See appendix, table 2A-1.)

Second, we told respondents, "Now, here are some concerns that people might have about the way these increased powers might be used by law enforcement. Would you say you have high concern, moderate concern, not much concern, or no concern at all about each of the following possibilities?" We then listed seven potential liberty-related concerns (appendix, table 2A-2).

Finally, we asked, "Overall, how confident do you feel that U.S. law enforcement will use its expanded surveillance powers in what you would see as a proper way, under the circumstances of terrorist threats?" Respondents could answer that they were very confident, somewhat confident, not very confident, or not confident at all. (See appendix, table 2A-3.)

Results of the September 2001 Survey

The survey conducted right after September 11 produced predictable but still striking results. All ten of the increased investigative powers for law enforcement in terrorist inquiries received majority support.

Six of the ten were favored by very large majorities—81 to 93 percent. The measures receiving the strongest support were

—stronger document and physical security checks for travelers (93 percent);

—expanded undercover activities to penetrate groups under suspicion (93 percent);

—stronger document and physical security checks for access to government and private office buildings (92 percent);

—use of facial recognition technology to scan for suspected terrorists at various locations and public events (86 percent);

—issuance of a secure ID technique for persons to access government and business computer systems (84 percent); and

—closer monitoring of banking and credit card transactions to trace funding sources (81 percent).

Adoption of a national ID system for all U.S. citizens was supported by 68 percent—a dramatic change from previous majority opposition to a general national ID or a national work identity card. In addition, 63 percent of the public favored expanded camera surveillance on streets and in public places and monitoring by law enforcement officials of discussions in chat rooms and other online forums. At the lowest but still majority level, 54 percent favored expanded government monitoring of cell phone conversations and e-mail.

Clearly, these results reflected the public's shock at the September 11 terrorist attacks. Traditional distrust of government powers was suspended in the belief that only the government could protect Americans from such outrages, and something like a wartime crisis gripped the nation at that moment.

However, in what I described at the time as a case of "rational ambivalence," the same respondents also voiced strong concern about the threats to civil liberties involved in the granting and exercise of expanded government powers. Majorities almost as large as those that approved these expanded powers—in the range of 68 percent to 79 percent—agreed that the use of these powers could pose a threat to individual liberties in the seven areas identified. (Table 2A-2 shows the responses to this question for 2001 and 2005.) Respondents in 2001 were concerned that

—judges who authorize investigations would not look closely enough at the justification of that surveillance (79 percent);

—Congress would not include adequate safeguards for civil liberties when authorizing these increased powers (78 percent);

—there would be broad profiling and searching of people based on their nationality, race, or religion (78 percent);

—the mail, telephone and cell phone calls, or e-mails of innocent people would be checked (72 percent);

—nonviolent critics of government policies would have their mail, telephone and cell phone calls, or e-mails checked (71 percent);

—law enforcement would investigate legitimate political and social groups (68 percent); and

—new surveillance powers would be used to investigate crimes other than terrorism (68 percent).

In other words, shortly after September 11, most of the public both approved of a sweeping set of expanded government investigative powers to fight terrorism and endorsed a set of concerns about the exercise of those powers.

In something of a tiebreaker, our third question probed this ambivalence by asking respondents how they balanced their fears about security and their worries about liberty at this moment in time. Again, the result was unequivocal. Eighty-seven percent of respondents in 2001 said they felt confident that law enforcement officials would use their expanded surveillance powers in what the respondent saw "as a proper way, under the circumstances of the terrorist threat." (Of this total, 34 percent were very confident and 53 percent somewhat confident.) Only 12 percent said they were not very confident or not confident at all. (Table 2A-3 shows the answers to this question for the first four surveys.)

Looking at the 2001 results, I noted at that time that the public's future views on these issues would depend on several developments after September 2001: whether there was another terrorist attack in the United states, how effectively law enforcement seemed to be using its new powers, whether legislatures and courts examined such uses and drew approved lines of control over any misuse, and how individuals themselves experienced the use of increased investigative powers.

Results of the June 2005 Survey

Over the next four years, what changed? Some expanded governmental powers saw drops in public approval. Support for a national ID system fell from 68 percent in 2001 to 61 percent in 2005—lower by 7 percentage points but still a majority. Law enforcement monitoring of the Internet had 57 percent support in 2005, down from 63 percent in 2001. And monitoring of cell phones and e-mail drew only 37 percent support, down 17 percentage points from the 54 percent recorded in 2001. However, support for the other government investigative or surveillance powers remained in the 59 to 81 percent range, down somewhat from the highs recorded in 2001 but still quite substantial.

At the same time, concern about the potential for violations of individual liberties remained at the same high 68 to 75 percent levels as in 2001. In other words, while approval of expanded government powers dropped somewhat in seven cases and slipped significantly in the remaining three over the next few years, the civil liberty concerns remained as salient in 2005 as in 2001.

How about the tiebreaker question? In 2004, 76 percent of the public said they were confident that law enforcement would use its expanded antiterrorist powers in a proper way. This was down from 87 percent in 2001. But it still represented a very solid three out of four Americans. When the "very" and "somewhat" confident responses are compared for the two years, the share of those who were somewhat confident was the same in 2001 and 2004—53 percent—but the share of those who were very confident dropped, from 34 percent in 2001 to 23 percent in 2004. (See table 2A-3.)

As in the past, confidence in the government's ability to use its expanded powers appropriately varied across groups. (See appendix, table 2A-4.) For example, conservatives were significantly more likely than liberals to express confidence in the government (87 to 55 percent), and white respondents were more likely to do so than black respondents (77 to 66 percent). As for the 24 percent of the overall public who did not share the majority view, it is useful to note that about 25 percent of respondents in privacy surveys from 1990 to 2002 said they had personally been the victim of what they felt was an improper invasion of their privacy.[9] And between 1995 and 2001, a similar 25 percent of the American public demonstrated intense attitudes toward consumer privacy in successive Harris-Westin surveys commissioned by Privacy and American Business, a think tank focusing on issues related to privacy and data protection.[10]

In June 2005 the tiebreaker question was rephrased since U.S. law enforcement was already using its expanded antiterrorist powers, and it was perceptions of these actual uses that we wanted to assess. The question asked, "Overall, thinking about the possibility of terrorist threats, do you feel that U.S. law enforcement is using its expanded surveillance powers in a proper way, or not?" In 2005, 57 percent responded that they felt these powers were being used in a proper way.

It is worth noting that this number might have been higher if the survey had taken place a few weeks later—in the wake of the subway and bus bombings that shook London in early July. Following these events, 55 percent of the public believed that similar attacks in the United States were likely, up from 35 percent approximately three weeks earlier.[11] As a result, support for antiterrorist measures probably also increased.

New Questions in the February 2004 and June 2005 Survey Series

We added two questions to the February 2004 and June 2005 surveys to probe the factors behind respondents' answers to our three primary trend inquiries. The results strengthened our confidence in the results of our tiebreaker query.

We recorded continued high overall approval of the Bush administration's antiterrorism record. When asked, "How would you rate the job that the Bush administration has done preventing a terrorist attack in the United States since September 11, 2001," 33 percent of respondents chose "excellent" and 37 percent "pretty good," for a 70 percent overall favorable rating in 2004. In 2005 there was a 13 percentage point decrease, with just 23 percent of respondents choosing "excellent" and 34 percent "pretty good," for a 57 percent overall favorable rating. Twenty percent chose "only fair" and 10 percent chose "poor" in 2004. In 2005, 23 percent chose "only fair," and 18 percent said "poor," an 11 percentage point increase overall in less than favorable rating.

Only 14 percent of the public said their own privacy had been significantly diminished as a result of government homeland security investigative programs. When asked, "How much do you feel government antiterrorist programs have taken your own privacy away since September 11, 2001," 35 percent said "none at all," and 25 percent said "only a little" in 2004. The 2005 figures show that 32 percent said "none at all," while "only a little" remained at 25 percent. Twenty-two percent said "a moderate amount" in 2004, while in 2005, 25 percent gave the same response. In 2004 only 14 percent felt that antiterrorism programs had had a significant impact on their own lives: 6 percent said "quite a lot," and 8 percent said "a great deal." In 2005 the share of respondents falling into this category rose by 3 percentage points: 7 percent said "quite a lot," and 10 percent said "a great deal."

Analysis and Implications of the Surveys

Clearly, the five trendline surveys reported here took a big picture approach. We asked about the acceptability of various types of government investigative powers, described in general terms. Then we asked about various liberty-related concerns, again in broad terms. We did not ask respondents whether they knew about or had opinions on specific homeland security programs, such as "know your customer" rules for opening financial accounts. Nor did we ask whether they believed other specific programs have followed acceptable privacy and due process rules or whether they believed personal data have been handled properly by government agencies.

This strategy reflected the fact that very few details were known in 2001 or even mid-2005 about how most homeland security investigative and surveillance activities were being conducted. Since the government rarely discloses the details of its security program operations, survey researchers cannot identify program-specific policy choices that members of the public are in a position to assess. With this caveat in mind, what conclusions can we draw from the 2001–2005 Harris-Westin trendline surveys?

Continuing High Support for Homeland Security Investigative Powers

Our 2005 survey delivered one clear message: 57 percent of the public expressed the view that law enforcement agencies are using their new or expanded investigative powers in a proper way, given the terrorist threat. This reading clearly stands apart from what the public may be feeling about larger post–September 11 issues, such as the failure of the Clinton and Bush administrations to anticipate and prevent the attacks and the need to revamp the nation's intelligence system to overcome conflicts between the FBI and CIA.

However, the overall reduction in approval levels for most of the government's expanded investigative powers shows that the impact of September 11 on public opinion has diminished somewhat after four years. Furthermore, the sharp drop in public support for e-mail and cell phone monitoring shows that respondents are more comfortable with some powers than with others. In 2005 many members of the public seemed concerned that e-mail and cell phone monitoring could be too sweeping and unfocused or that such unsupervised actions might not really be needed to fight or deter terrorists.

Taken as a whole, the continuing public approval of antiterrorist programs registered in these surveys contrasts with the attacks mounted by liberals, some libertarian conservatives, and political critics of the Bush administration. There is strong evidence of continuity with the historic tradition of public support for law enforcement and security officials when the nation appears to face genuine and grave dangers. Between 2001 and 2005, a large majority clearly believed that it did not feel it had the luxury of exhibiting the traditional American distrust of government, which predominates in nonemergency times.

Continuing Concern about Liberty Safeguards

Our surveys also recorded concern that limitations and safeguards may not be applied properly in the execution or oversight of investigative programs. The fact that these liberty concerns did not diminish between 2001 and 2005 suggests that the public expects the White House, Congress, and the courts to install meaningful liberty protections in antiterrorist programs. Further evidence pointing in the same direction comes from the wholesale rejection in 2002–03 of the Defense Department's Total Information Awareness program. This episode showed the firestorm that can arise in response to policies that are seen as poorly conceived, ineptly organized, and lacking adequate concern for civil liberties.

Late 2005 and Early 2006 Developments

The security-liberty scene was dramatically altered in late 2005 when news reports disclosed that the Bush administration had authorized the National Security Agency to conduct secret and judicially unwarranted monitoring of telephone calls and e-mails to and from the United States by terrorist suspects.

This set off a sharp debate in Congress and the media over the legality of this program, especially the administration's decision to avoid the special foreign intelligence court created to review and issue national security wiretap orders and to rely only on the president's general executive authority.

How the public perceived this activity, and how it affected overall public attitudes toward the privacy versus security balance, was probed in early January 2006 by a poll conducted by ABC News and the *Washington Post*.[12] The key results showed:

—66 percent said they were following the news stories about the NSA monitoring program.

—51 percent said they considered "this wiretapping of telephone calls and e-mails without court approval" to be "acceptable," while 47 percent said it was "unacceptable."

Assessing the larger security and privacy balance, the poll updated its trend questions as follows:

"What do you think is more important right now—for the federal government to investigate possible terrorist threats, even if it intrudes on personal privacy, or for the federal government not to intrude on personal privacy, even if that limits its ability to investigate possible terrorist threats?"

To this question, 65 percent chose "investigate threats" and 32 percent responded "respect privacy."

This 2006 reading in favor of investigation was down from 73 percent in 2003 and 79 percent in 2002. Likewise, those choosing "respect privacy"—at 32 percent—increased from 21 percent in 2003 and 18 percent in 2002.

To another question asked respondents (whether—in investigating terrorism—they think "federal agencies are or are not intruding on some Americans' privacy rights"), 64 percent now believe such intrusions are taking place, up from 58 percent in 2003

When next asked whether such intrusions are or are not justified, 49 percent said yes, they are justified (down from 63 percent in 2003), while 46 percent said not justified (up from 29 percent in 2003).

In another question, respondents were asked which worried them more—"that Bush will not go far enough to investigate terrorism because of concerns about constitutional rights or that Bush will go too far in compromising constitutional rights in order to investigate terrorism?" Forty-eight percent worry the President will not go far enough but 44 percent worry he will go too far (6 percent chose neither option and 2 percent no opinion).

At the overall level, 53 percent say they now approve of the way President Bush "is handling the U.S. campaign against terrorism" (down from 56 percent in 2005), while 45 percent say they disapprove (up from 44 percent in 2005).

This early 2006 reading of public attitudes, in my judgment, supports and even strengthens the judgments presented in the main conclusion section—that retaining public confidence in the anti-terrorist programs of the administration now demands the institutionalization of meaningful privacy and due process safeguards.

Conclusion

We are entering phase two of the security-versus-liberty debate. This will be a period of reorganizing for the long haul. The White House now faces the challenge of designing and incorporating meaningful liberty and due process systems in all major homeland security operations, without compromising their effectiveness.

The concerns and ambivalence registered by the public on the security-versus-liberty dimensions of homeland security remain. Rather than a challenge, the administration should see these public opinion trends as an opportunity. By providing appropriate responses to these concerns, the government can help ensure that public support for antiterrorist programs remains strong and that our security programs incorporate the civil liberty processes that define our fundamental national identity.

Notes

1. Before looking at the data, it may be helpful to note briefly the strengths and weaknesses of survey methodology. When compared to declarations by interest groups, political leaders, and the media that "we speak for the people and know what they want or feel," surveys potentially offer a more objective and structured reading of public attitudes. However, the quality of surveys depends on many subjective factors: the wording of questions and responses; the placement and order of questions; the timing of the survey in relation to major events; the quality of the sample; the survey methodology (telephone, online, in person); any weighting of responses done by the survey firm; the nonresponse rate and its effects; and the credibility of the survey summary and analysis. In addition, it always helps to note who is sponsoring the survey, to consider how well the underlying policy issues on a subject have been captured and presented to respondents, and to track how any new survey compares with previous surveys on the same topic.

2. Pew Internet and American Life Project, *Fear of Online Crime: Americans Support FBI Interception of Criminal Suspects' Email and New Laws to Protect Online Privacy* (Washington, D.C., April 2001). For this survey Princeton Survey Research Associates interviewed 2,096 respondents via telephone from February 1 through March 1, 2001. The margin of error was ± 2 percentage points.

3. Louis Harris and Associates and Alan F. Westin, *Equifax-Harris Mid-Decade Consumer Privacy Survey 1995* (New York: Louis Harris and Associates,

1995). This survey conducted for Equifax, Inc., entailed telephone interviews of 1,006 adults of the national public from July 5 though July 17, 1995.

4. Pew, *Fear of Online Crime.*

5. Louis Harris and Associates and Alan F. Westin, *The Equifax Report on Consumers in the Information Age* (New York: Louis Harris and Associates, 1990). In this survey conducted for Equifax, Inc., 2,254 respondents of the national public were interviewed by telephone during 1990.

6. Louis Harris and Associates and Alan F. Westin, *Equifax-Harris Consumer Privacy Survey 1994* (New York: Louis Harris and Associates, 1994). This survey conducted for Equifax, Inc., entailed telephone interviews of 1,005 adults in a national sample during 1994.

7. Opinion Research Corporation and Alan F. Westin, *Public Records and the Responsible Use of Information* (Hackensack, N.J.: Privacy and American Business, November 2000). In this survey, commissioned by ChoicePoint, 1,011 respondents were interviewed via telephone from September 26 through October 6, 2000. The margin of error was ± 3 percentage points.

8. For the September 2001 survey, see Harris Interactive, "Overwhelming Public Support for Increasing Surveillance Powers and, In Spite of Many Concerns about Potential Abuses, Confidence that These Powers Would Be Used Properly," *Harris Poll,* no. 49, October 3, 2001 (www.harrisinteractive.com/harris_poll/index.asp?PID=260). For the March 2002 survey, see Harris Interactive, "Homeland Security: Public Continues to Endorse a Broad Range of Surveillance Powers But Support Has Declined Somewhat Since Last September," *Harris Poll,* no.16, April 3, 2002 (www.harrisinteractive.com/harris_poll/index.asp?PID=293). For the February 2002 survey, see Harris Interactive, "Homeland Security: American Public Continues to Endorse a Broad Range of Proposals for Stronger Surveillance Powers, but Support Has Declined Somewhat," *Harris Poll,* no. 14, March 10, 2003 (www.harrisinteractive.com/harris_poll/index.asp?PID=362). For the February 2004 survey, see Harris Interactive, "Strong and Continuing Support for Tough Measures to Prevent Terrorism," *Harris Poll,* no. 17, March 5, 2004 (www.harrisinteractive.com/harris_poll/index.asp?PID=446). For the June 2005 survey, see Harris Interactive, "Majority Believes U.S. Law Enforcement Is Using Its Expanded Investigative Powers Properly in Terrorist Cases but Over Two-Thirds Worry about Insufficient Civil Liberty Safeguards," *Harris Poll,* no. 51, June 29, 2005 (www.harrisinteractive.com/harris_poll/index.asp?PID=580).

9. Harris-Westin privacy surveys conducted from 1990 through 2002. For specific survey information, see www.pandab.org.

10. Questions from the Harris-Westin privacy surveys conducted between 1995 and 2005 demonstrated the Westin Privacy Segmentation Index. For more information about this index, contact Irene Oujo at Privacy and American Business, Hackensack, New Jersey.

11. Gallup Organization, *Perceived Terrorist Threat in U.S. Rises after London Bombing,* CNN–USA Today–Gallup Poll (Princeton, N.J., July 2005). Survey entailed telephone interviews with 1,006 adult Americans during 2005.

12. *Broader Concern on Privacy Rights but Terrorism Threat Still Trumps,* ABC News/*Washington Post* Poll, January 10, 2006.

Appendix 2A: Results of Harris-Westin Surveys, September 2001 to June 2005

Table 2A-1. Favor versus Oppose Ten Proposals for Increased Law Enforcement Powers[a]

Percent

Proposal	Survey date	Favor	Oppose	Not sure/ decline to answer
Stronger document and physical security checks for travelers	Jun. 2005	81	17	2
	Feb. 2004	84	14	1
	Feb. 2003	84	14	1
	Mar. 2002	89	9	2
	Sept. 2001	93	6	1
Stronger document and physical security checks for access to government and private office buildings[b]	Feb. 2004	85	14	1
	Feb. 2003	82	15	2
	Mar. 2002	89	10	1
	Sept. 2001	92	7	1
Expanded undercover activities to penetrate groups under suspicion	Jun. 2005	76	20	4
	Feb. 2004	80	17	3
	Feb. 2003	81	17	2
	Mar. 2002	88	10	2
	Sept. 2001	93	5	1
Use of facial recognition technology to scan for suspected terrorists at various locations and public events[b]	Feb. 2004	80	17	3
	Feb. 2003	77	20	3
	Mar. 2002	81	17	2
	Sept. 2001	86	11	2
Issuance of a secure ID technique for persons to access government and business computer systems, to avoid disruptionsb	Feb. 2004	76	19	5
	Feb. 2003	75	21	4
	Mar. 2002	78	16	6
	Sept. 2001	84	11	4
Closer monitoring of banking and credit card transactions, to trace funding sources	Jun. 2005	62	35	3
	Feb. 2004	64	34	3
	Feb. 2003	67	30	2
	Mar. 2002	72	25	2
	Sept. 2001	81	17	2
Adoption of a national ID system for all U.S. citizens	Jun. 2005	61	34	5
	Feb. 2004	56	37	2
	Feb. 2003	64	31	5
	Mar. 2002	59	37	5
	Sept. 2001	68	28	4

Table 2A-1. Favor versus Oppose Ten Proposals for Increased Law
Enforcement Powers[a] (continued)
Percent

Proposal	Survey date	Favor	Oppose	Not sure/ decline to answer
Expanded camera surveillance on streets and in public places	Jun. 2005	59	40	1
	Feb. 2004	61	37	2
	Feb. 2003	61	37	1
	Mar. 2002	58	40	2
	Sept. 2001	63	35	2
Law enforcement monitoring of Internet discussions in chat rooms and other forums	Jun. 2005	57	40	4
	Feb. 2004	50	45	6
	Feb. 2003	54	42	4
	Mar. 2002	55	41	4
	Sept. 2001	63	32	5
Expanded government monitor- ing of cell phones and e-mail, to intercept communications	Jun. 2005	37	60	3
	Feb. 2004	36	60	4
	Feb. 2003	44	53	4
	Mar. 2002	44	51	4
	Sept. 2001	54	41	4

Source: Harris Interactive, *Harris Poll*, no. 49, October 3, 2001; no. 16, April 3, 2002; no. 14, March 10, 2003; no. 17, March 5, 2004; and no. 51, June 29, 2005.

a. Base: all adults. Survey statement read: "Here are some increased powers of investigation that law enforcement agencies might use when dealing with people suspected of terrorist activity, which would also affect our civil liberties. For each, please say if you would favor or oppose it." Note that percentages may not add up exactly due to rounding.

b. Question not repeated in June 2005 survey.

Table 2A-2. Levels of Concern about Seven Potential Abuses of Power[a]

Percent

Potential abuses	Survey date	High	Moderate	Not much	None	Not sure/ decline to answer	Total concern (high + moderate)
Judges who authorize investigations would not look closely enough at the justification of that surveillance.	Jun. 2005	37	38	16	7	2	75
	Oct. 2001	44	35	11	7	2	79
Congress would not include adequate safeguards for civil liberties when authorizing these increased powers.	Jun. 2005	38	37	14	8	2	75
	Oct. 2001	39	39	12	8	2	78
There would be broad profiling of people and searching them based on their nationality, race or religion.[b]	Feb. 2004	42	31	13	13	1	73
	Oct. 2001	44	33	11	10	1	78
The mail, telephone, e-mails, or cell phone calls of innocent people would be checked.[b]	Feb. 2004	47	29	13	10	1	76
	Oct. 2001	45	27	13	14	1	72
Nonviolent critics of government policies would have their mail, telephone, e-mails or cell phone calls checked.[b]	Feb. 2004	40	36	13	11	1	76
	Oct. 2001	38	33	14	14	<0.5	71
Law enforcement would investigate legitimate political and social groups.	Jun. 2005	29	39	20	11	2	68
	Oct. 2001	32	36	16	15	1	68
New surveillance powers would be used to investigate crimes other than terrorism.[b]	Feb. 2004	35	36	16	11	1	71
	Oct. 2001	32	35	15	16	1	68

Source: See footnote to table 2A-1.

a. Base: all respondents. Survey question read: "Now, here are some concerns that people might have about the way these increased powers might be used by law enforcement. Would you say you have high concern, moderate concern, not much concern, or no concern at all about each of the following possibilities?" Note that percentages may not add up exactly due to rounding.

b. Question not repeated in June 2005 survey.

Table 2A-3. Confidence That Surveillance Powers Used in Proper Way[a]
Percent

Response	September 2001	March 2002	February 2003	February 2004
Very confident	34	12	22	23
Somewhat confident	53	61	52	53
Not very confident	8	17	14	15
Not confident at all	4	6	9	9
Not sure/decline to answer	1	3	2	1

Source: See footnote to table 2A-1.

a. Base: all adults. Survey question read : "Overall, how confident do you feel that U.S. law enforcement will use its expanded surveillance powers in what you would see as a proper way, under the circumstances of terrorist threats? Would you say very confident, somewhat confident, not very confident, not confident at all?" In June 2005 the question was rephrased: "Overall, thinking about the possibility of terrorist threats, do you feel that U.S. law enforcement is using its expanded surveillance powers in a proper way, or not?" Note that percentages may not add up exactly due to rounding.

Table 2A-4. Demographic Variations, Confidence in Proper Use, February 2004
Percent

Category	Very plus somewhat confident
Overall public	76
By political philosophy	
Conservative	87
Moderate	78
Liberal	55
By race	
Hispanic	79
White	77
Black	66
By gender	
Female	80
Male	72
By age, selected	
65+ years	80
25–29 years	70
By party identification	
Republican	92
Independent	72
Democrat	69
By education	
High school or less	78
College/postgraduate	68

Source: See footnote to table 2A-1.

II

Protecting Security and Liberty:
Information Technology's Role

3

Information Technology and the New Security Challenges

JAMES STEINBERG

On September 11, 2001, nineteen men affiliated with the al Qaeda terrorist organization hijacked four U.S. airliners and inflicted the most serious "surprise attack" on the United States since the infamous attack on Pearl Harbor nearly sixty years earlier. Just as in the aftermath of Pearl Harbor, many began to question whether the United States should have been surprised—and even whether the country should have been able to thwart the attack.

Certainly, there was no "strategic" surprise. Al Qaeda's leaders had explicitly declared their intention to harm the United States and had carried out attacks against Americans abroad. In addition, there were reasons to suggest that the *type* of attack could have been known in advance. The World Trade Center had been the object of terrorism eight years before. Individuals associated with al Qaeda had previously threatened to crash airplanes into the headquarters of the Central Intelligence Agency (CIA) and the Eiffel Tower. Two different Federal Bureau of Investigation (FBI) offices had raised concerns about suspicious activities at flight training schools in the United States.

But even more important, it is clear, at least in hindsight, that it might have been possible to discover the identity of the hijackers in advance.

Two of the hijackers were on CIA and State Department watch lists because they had met with senior al Qaeda leaders in Malaysia. They had bought their airline tickets using their real names. From those tickets it would have been possible to discover their home addresses, which in turn would have revealed the identities of three more of the hijackers who shared that address. There were traceable links between these five and the remaining hijackers. In an "after the fact" reconstruction, all of these connections were revealed using a relatively simple software solution originally developed to detect collusion at casinos.

The fact that the plot might have been thwarted does not by itself make the case for blame. "Could have" does not mean "should have" been able to identify and prevent the attack. Rather, this tale serves as a cautionary lesson and highlights the crucial role of information and information technology in dealing with the new security challenges facing the United States in the twenty-first century. The bad news in this story is that information was available that might have enabled us to thwart the attack, but we failed to mobilize the information effectively so that the right people had that information in a timely, useable way. The good news is that tools are available that will allow us to use information far more effectively in the future to protect our security.

How can information technology and information strategies be used to tackle these new, dangerous threats? What unique challenges—and opportunities—does the use of these technologies present for cooperation between the government and the private sector? To answer these questions, it is essential to understand how this challenge differs from what the United States has had to deal with in the past. This understanding, in turn, will allow us to evaluate some promising strategies that can be pursued in the future.

The Challenge

The intelligence challenge has four key components. The first is collecting timely, relevant, and, in the best case, actionable information—"news you can use." The second component is collating or bringing together information from the full spectrum of sources. The third is analyzing the information—"connecting the dots." The final component is disseminating that information to those who need to act on it—such as policymakers, law enforcement officials, people charged with the security of critical infrastructure, and the public—in a form that allows them to use that information to accomplish their mission.

In the fight against terrorism, accomplishing each of these tasks is far more difficult than it was during the cold war. At that time the United States faced a known enemy. The United States generally knew what to look for and where to look for it, although it had to deal with its adversaries' attempts to conceal crucial information. For the most part, the information needed was overseas and largely concerned military activities. This limited the need to collect information in the United States or about U.S. citizens. Most of the expertise and knowledge required to analyze the collected information resided in the federal government. Finally, the actions that had to be taken rarely involved the public, the private sector, or state and local officials.

All of this has now changed. Today, terrorists threaten the United States both at home and abroad. They have no fixed address, and only rarely are their names or their targets known. Technology and globalization have made it easier for would-be terrorists to bring dangerous people and weapons into the United States and to conceal their activities. Key information needed to detect and prevent terrorist attacks lies in the private sector—at airlines and flight schools, with operators of chemical plants and high-rise buildings, with community doctors. Increasingly, the private sector must be counted on to take the actions necessary to prevent attacks or deal with their consequences. For all these reasons, the United States must adapt its intelligence efforts and the organization of its so-called intelligence community to meet this radically different challenge.

How has the United States done so far in meeting the challenge? Governments at all levels have taken significant steps since September 11 to heighten their focus on these new threats. At the federal level, Congress has expanded the authority of the federal government to collect and share information through the USA PATRIOT Act of 2001, which has broken down the rigid wall that separated the law enforcement and intelligence communities. Congress also created the Department of Homeland Security, which has wide-ranging responsibilities for border protection, immigration, supporting local response to terrorist attacks, and protecting critical infrastructure. It also has a special responsibility to build ties with state and local governments and the private sector. In addition, the department has an office charged with developing new technologies to meet the threat—a kind of domestic equivalent of the Defense Advanced Research Projects Agency (better known as DARPA);

its mission is to cut through the traditional procurement red tape to bring innovative ideas into service quickly. Furthermore, new funding has been provided for first responders—local officials such as police, fire-fighters, and public health workers. The Pentagon even set up a new military command—Northcom—designed to coordinate the military side of homeland security.

More recently, spurred by the 9/11 Commission's recommendations in its report of July 2004, Congress passed legislation, subsequently signed by the president, to reorganize intelligence activities and organizations.[1] This law created the position of Director of National Intelligence (DNI), with power over budgets and personnel throughout the intelligence community (both foreign and domestic). It also established a multiagency National Counterterrorism Center (with the director of the NCTC reporting to the president and the DNI) to coordinate analysis and joint operational planning, and set up a civil liberties board. The president also issued executive orders for immediate development of common standards for the sharing of terrorism information within and among the various intelligence and counterterrorism agencies and related authorities in state and local governments. In addition, the orders called for the creation of a council, chaired by the Office of Management and Budget, to plan for and oversee the establishment of automated terrorism information sharing among appropriate agencies.

Thinking Fresh

These are important steps, but much still needs to be done. To successfully meet the new security challenges of the future, the U.S. approach must be reconstructed from the bottom up. This entails rethinking all four core elements of the information challenge: what information to collect, how to bring the disparate pieces together, how to analyze the information, and how to share it.

The United States must begin by recognizing the limits of relying on what one could call the "known enemy" or "known perpetrator" model. This is the approach that characterized the CIA's and FBI's thinking during the cold war years. Of course, sometimes the government will know who to go after and can focus its intelligence collection on a particular individual or group—that is how Khalid Sheikh Mohammed, al Qaeda's chief of operations, was tracked down and captured. Secretary of Defense

Donald Rumsfeld has called these the "known unknowns," and they pose the most straightforward intelligence challenge.

But in many cases neither the identity of the adversaries nor their intended targets will be known. Such a situation requires that the capacity for a wide scan—or peripheral vision—be developed to reduce the chance of surprise. The goal of this hunt is to find what Rumsfeld calls the "unknown unknowns." The national security system must be able to detect anomalous patterns that warrant investigation, identify critical vulnerabilities that might tempt would-be terrorists, and identify the materials that an attacker might use to craft weapons of mass harm. In general, ways must be found to ensure that the U.S. is not fighting the last war or thwarting the last attack.

One of the most thoughtful intelligence analysts of modern times— Richards Heuer—has written extensively about how to avoid what he called mental ruts, the "grooves" in the neural pathways that lead individuals to organize new data into familiar categories.[2] Today, new information technologies, such as data mining and pattern recognition, and analytical tools can help in avoiding this pitfall. One example is software that helps analysts create new taxonomies based on clustering of unstructured information without prespecifying the categories—a way of overriding mental ruts by enforcing open-mindedness.

A broad scan demands that a wide variety of data be examined. Unlike the days of the cold war, the government does not own much of the data that it will want to exploit. Instead they exist in the private sector—in airline reservation systems, credit card company records, and Internet service providers' logs. Used wisely, these information sources can provide an invaluable supplement to data collection done directly by law enforcement and intelligence agencies. For example, one data management tool developed by the private sector uses an online database that cross-references billions of current and historical records; it has been used to assist in criminal investigations, such as the Washington-area sniper case in 2002. Public health officials concerned about bioterrorism (not to mention natural epidemics) are also learning to tap private sector information, such as over-the-counter drug sales, as an important surveillance tool.

Of course, these tools can be a double-edged sword: they offer unprecedented ability to collect and analyze data but also raise fundamental questions about privacy and civil liberties. These tensions could be seen in the dispute over the Pentagon's proposed Total Information Awareness (renamed the Terrorism Information Awareness) program.

The program's advocates proposed harnessing modern search technology to access billions of bits of data in privately held databases—without a search warrant. The ensuing uproar caused the program to be shelved.

The way federal entities collected and used information during the 1960s and early 1970s has made many Americans leery about sharing private information with the government. But in today's environment, most Americans recognize that new methods must be developed to deal with new threats. That is why what is needed is an approach that can reconcile civil liberties with security. Fortunately, information technology can be a powerful tool for meeting this challenge as well.

To mitigate many of the concerns, new, clear policy guidelines must be developed to define under what circumstances the government should be able to access information from the private sector. By constraining officials in their use of private information, such guidelines would help protect individual rights and safeguard the interests of companies holding the information. At the same time, they would also empower government officials to share information by clarifying what is permissible. But these rules will only be useful if they can be enforced. Thus it is critically important to develop accountability tools that audit the actions of both government and the private sector in acquiring and sharing sensitive personal information. These audit tools can be built directly into the software and middleware used to acquire, store, and distribute information. For example, although there was much wrong with the FBI's controversial "Carnivore" software, which allowed the government to monitor communications over the Internet, one of its valuable features was an automatic audit capability that allowed for subsequent review of any e-mails selected by Carnivore. This created a record that could be examined by federal judges who could determine whether the system was being used in a manner consistent with the court order that authorized its use.

Americans will have to reach a new understanding of their responsibilities, as well as their rights, as private citizens in providing information to the government. In Washington there was an outcry when the Justice Department, through the so-called Operation TIPS (Terrorist Information and Prevention System), attempted to recruit letter carriers and meter readers to report "suspicious" information that might be relevant to terrorism. There was so much criticism that Congress specifically banned the program. Yet for a number of years, Congress and the administration have urged owners of critical private sector cyberinfrastructure to report suspicious activities, with few outcries from the public. In the war on terrorism, the private sector will need to play a bigger role, but guidelines and accountability will be crucial if the public is going to have confidence in wider private-sector participation.

Another key task—in addition to widening our scan—is to construct an entirely new approach to sharing information and analysis. During the cold war, the United States developed a very rigid system designed to protect intelligence secrets from the enemy. People with access to sensitive information were required to obtain security clearances based on extensive background checks. Then, in addition to the overall security clearance, access was further limited by requiring a specific, predetermined "need to know."

By putting a premium on security, this system protected against unauthorized disclosures. However, its inherent rigidity made it difficult to share information even *within* individual government agencies, much less with other federal departments, state and local governments, or especially with the private sector. Indeed, the joint Senate-House Intelligence Committee that investigated the September 11 attacks identified faulty information sharing as being at the heart of the problem.

> Within the Intelligence Community, agencies did not adequately share relevant counterterrorism information, prior to September 11. This breakdown in communications was the result of a number of factors, including differences in the agencies' missions, legal authorities, and cultures. Information was not sufficiently shared, not only between different Intelligence Community agencies, but also within individual agencies, and between the intelligence and the law enforcement agencies. . . . Serious problems in information sharing also persisted, prior to September 11, between the Intelligence Community and relevant non-Intelligence Community agencies. This included other federal agencies as well as state and local authorities. This lack of communication and collaboration deprived those other entities, as well as the Intelligence Community, of access to potentially valuable information in the "war" against Bin Ladin.[3]

The now famous "Phoenix" FBI memo illustrates clearly the nature of the problem. A field agent had what turned out to be a brilliant hunch.[4] But that memo never made it to the attention of agents in Minnesota who were dealing with the case of Zacarias Moussaoui—the alleged hijacker who wanted to learn to fly a plane but not land it.

Meeting the new threat also requires a different approach to information sharing, based on two basic components. The first of these is a decentralized network that does not try to fix its membership in advance. Who could have known before September 11 that flight school instructors in Minnesota and Florida could provide crucial data to help track down potential terrorists? Fortunately, technology combined with a new mindset can help address this challenge through tools such as software that helps route communications to individuals who, based on their past activities and interests, are likely to be valuable recipients. The sender does not have to fill in the "to" line on the routing slip, and the system continually updates itself based on the ongoing workflow of people in the network. The lessons of peer-to-peer computing also offer important insights into how to build effective networks without excessive reliance on centralized control.

The public health network that spontaneously arose to identify and track the virus that causes severe acute respiratory syndrome (SARS) illustrates the kind of self-forming community that will be necessary to tackle these new threats.[5] Dr. Klaus Stohr, manager of the World Health Organization's flu program, was attending a routine flu vaccine meeting in Beijing when a healthcare worker from Guangdong told him that several people in his region had died from an unusually severe flu in November 2002. As more reports of a mysterious flu appeared in February and early March 2003, Dr. Stohr set to work. He learned that labs in Canada, Vietnam, and Germany had samples of the virus, but he thought it would be desirable to expand the network. Drawing on the World Health Organization's global surveillance network, he persuaded a total of twelve labs to join in the search. Within days labs in Hong Kong and Germany had electron microscope pictures of the virus, which were immediately flashed on the Internet. A Centers for Disease Control and Prevention (CDC) lab in Atlanta confirmed that a similar looking virus drawn from a SARS patient was capable of killing monkey cells—a strong clue to the virus's lethality.

To determine whether this was a previously known virus, the CDC mailed a sample to a lab at the University of California, San Francisco, which had pioneered a virus-detecting silicon "gene chip" containing fragments from all known viruses. The lab quickly confirmed the finding of a corona virus, but of a type not seen before. This in turn helped jumpstart the CDC's effort to sequence the new virus's genome. A separate team in Vancouver, Canada, also began to sequence the genome, placing nearly half of the lab's staff on the project. Within a few days, both labs had completed the sequencing of the genome, posting their results on the web for researchers around the world, who immediately set to work on creating a vaccine. In a matter of several weeks, this informal network of researchers and clinicians had nailed the culprit. Compare this to twenty years earlier, when it took two and a half years to isolate the virus that causes AIDS.

To be successful, the United States cannot expect to build such a system from scratch or to build a system that will last for all time. Technologies such as XML (extensible markup language) should be used to allow communication to transcend legacy systems. Open standards are necessary to ensure interoperability without the rigidity of prescribed standards that stifle innovation. The concept must be a "network of networks." And it must be useful for a broad variety of applications. It simply makes no sense to spend billions of dollars for a very, very unlikely probability—say the possibility of a bioterror attack in Wichita, Kansas. But if the same information sharing and response system can help deal with naturally occurring diseases or an accidental chemical spill or even a local crime wave, then the cost-benefit calculation will be much more favorable. And such a multipurpose system will give users a greater incentive to actually use it.

This approach will help ensure that data that might be relevant to a terrorist attack are fed into the network without the collector of these data having to make a conscious decision to define an event or observation as "terrorism related." An example from the world of cybersecurity illustrates very clearly the importance of not prematurely trying to figure out which "box" to put a problem in. Just a few years ago, the infamous Love Bug hit the world's servers, causing billions of dollars in damage and by one estimate affecting 80 percent of federal agencies, including both the Defense and State Departments. Was it a technical malfunction, a mischievous adolescent hacker, a terrorist group, or a foreign government adversary? No one knew at first. Who should be in charge of stopping it? No one had the luxury of time to engage in turf fights. Within minutes of the first reported attacks, an informal network of computer security specialists, corporate managers, and government officials at the National Infrastructure Protection Center began to respond both to halt the spread of the virus and to pinpoint the origin of the attack.

To complement the development of a decentralized network, a second component is needed: a fresh way to reconcile effective information sharing with protecting sensitive information. The current system is highly compartmentalized. As mentioned earlier, the current system requires people to get security clearances—confidential, secret, or top secret—to simply enter the game. Then, on top of that, access to each bit of sensitive information is confined to a very small percentage of those who have the relevant level of security clearance, with very few given access to a large number of "compartments." This approach limits the potential damage if an untrustworthy person gets through the screening system of security clearances. But it also means that very few people see the big picture. It is like the classic story of people trying to describe an elephant in the dark even if they can only touch one part of the elephant's body. The compartmentalized approach also means one needs to know in advance who is entitled to get the information—something that may not be knowable until it is too late.

Part of this problem can be solved with new technology. Software tools today allow a classified document to be shared by e-mail, with recipients only having access to those portions of the document for which they have clearance. These programs provide the verification and authentication tools that would allow multiple users with different levels of security clearance to work off a common document.

However, some of the answer must also come through accepting a different trade-off between openness and security. More widespread information sharing might expose new vulnerabilities that terrorists could exploit, or it might help them elude capture by tipping them off. But, as the open source code experience has shown, it could also lead to quicker, more effective action to eliminate the vulnerabilities once discovered. And sometimes averting a terrorist attack is more important than capturing the perpetrator after the fact. The alternative mindset—one that led government agencies to remove thousands of web pages from public circulation after September 11—may have some short-term benefits. But reversion to a "secrecy first" model would not necessarily make this country more secure and might imperil the very freedoms and government accountability that we are trying to defend.

These twin challenges—creating a decentralized network and removing the barriers to information flow—were very much in play as the administration set out to create the Terrorist Threat Integration Center (TTIC) in 2003. The idea behind the TTIC was very much a product of the perceived deficiencies that existed before September 11—in particular, the lack of a single place where all the relevant data could be fused and made available to those who needed them. But the TTIC was unable to overcome the problems that plagued its predecessor (the CIA's Counterterrorist Center) and was not very effective at linking the people who had the requisite information to those who needed to use it. Analysts from various agencies sat in a common room with access to their "home" agencies' intelligence, which could be shared manually among analysts at the TTIC but not beyond without permission from the home agencies—and the TTIC itself was separated by physical barriers from access by other agencies. The TTIC has since been superseded by the National Counterterrorism Center, which provides more centralized authority over both analysis and operational planning, although not over counterterrorism operations themselves.

The discussion above serves as a reminder that decentralization has its limits. It is all well and good if a system can be structured so that the suspicions of an FBI field agent in Phoenix get communicated to the airport official with a suspicious customer in Minnesota. But somebody still needs to make the decision to do something about it. The network, therefore, also needs awareness and accountability.

The government can introduce new technologies and reorganize the way that government agencies operate, as the September 11 Commission report pointed out, "human" resistance is as much an impediment to information sharing as "systemic" resistance. Therefore, the culture within these agencies must change as well. There is a need for policies that will address the cultural and bureaucratic disincentives to sharing information and encourage a mindset of "writing to share." Elements of such policies could include career incentives to participate in "joint" assignments, joint training and education, and assignments to billets in other services.

Conclusion

The challenge ahead is daunting. The terrorist only needs to get it right once in a thousand tries to be successful. If the defenders fail to pick up a clue even once in a thousand times, the results can be catastrophic. The security of the United States will depend in large measure on whether it is smarter than its adversaries. And that, in turn, depends on whether ways can be found for government at all levels and the private sector to harness the power of information for the public good.

Notes

1. P.L. 108-458.

2. Richards Heuer, *Psychology of Intelligence Analysis* (CIA, 1999).

3. Senate Select Committee on Intelligence and House Permanent Select Committee on Intelligence, "Part 1: The Joint Inquiry. Findings and Conclusions," *Joint Inquiry into Intelligence Community Activities before and after the Terrorist Attacks of September 11, 2001*, H. Rept. 107-792 and S. Rept. 107-351, 107 Cong. 2 sess. (Government Printing Office, December 2002).

4. A couple of months before the September 11 attacks, FBI special agent Kenneth Williams sent a memo "to advise the Bureau and New York of the possibility of a coordinated effort by Usama Bin Laden (UBL) to send students to the United States to attend civil aviation universities and colleges. . . . these individuals will be in a position in the future to conduct terror activity against civil aviation targets." This emerged from the hearings by the Senate Select Committee on Intelligence and House Permanent Select Committee on Intelligence.

5. For details see Elena Cherney and others, "Cellular Sleuths: How Global Effort Found SARS Virus in Matter of Weeks," *Wall Street Journal*, April 16, 2003, p. A1.

4

Building a Trusted Information-Sharing Environment

ZOË BAIRD AND JAMES BARKSDALE

O ur nation stands at a critical juncture in the relationship of our government to our people. In the wake of the September 11 attacks, the threat of terrorism has raised the possibility that individuals, rather than states, pose the primary threat to our security. As a result, the government requires more information about individuals at home and abroad than ever before—a requirement that poses a challenge to our laws and values. To overcome this challenge, we will need to realign the ability to inquire into individuals' activities without undermining the traditional balance between the rights of individuals and the needs of the government.

Making this challenge even more difficult is the fact that traditional intelligence and law enforcement approaches to protecting our nation from terrorism are no longer adequate. For example, every potential target cannot possibly be hardened; even when soft targets, buildings, or airports are secured, they remain vulnerable to attack. In order to combat terrorism, we need to rely on a strategy of prevention. This strategy, in turn, relies on good intelligence. Information is critical in this battle as it

The authors would like to thank Mary DeRosa, Todd Glass, Tara Lemmey, and Stefaan Verhulst for their contributions to this article.

allows us to stop attacks before they occur. Good information also allows us to use our resources more efficiently, focusing their application in areas where terrorism is most likely to strike.

In December 2003 the Markle Task Force on National Security in the Information Age (subsequently referred to as the Task Force), a distinguished panel of security experts from five administrations, as well as experts on technology and civil liberties, released its second report, *Creating a Trusted Information Network for Homeland Security*.[1] This report, along with its 2002 predecessor, focused on the role of information and information technology in combating terrorism.[2] Arguing that information technology is critical to intelligence sharing—which, as we now know, was one of the key failures of pre–September 11 counterterrorism efforts—the Task Force called for the adoption of nationwide intelligence network capabilities to help intelligence and law enforcement agencies "connect the dots" across regions and jurisdictions and draw a meaningful picture from apparently dispersed bits of data. This proposed system was named the System-wide Homeland Analysis and Resource Exchange (SHARE) Network.

The September 11 Commission endorsed the SHARE concept in its final report, and the president and Congress subsequently took a number of steps that will help bring a "trusted information environment" into being that can achieve the goals developed by the Task Force. A set of August 2004 executive orders, as well as the Intelligence Reform and Terrorism Prevention Act of 2004, have set the country firmly on the path to a more effective information-sharing environment.[3] However, these developments represent only the first milestone on the path to real reform. The manner in which the executive orders and the act are implemented will determine the success of intelligence reform and, ultimately, our national security. Therefore, we also discuss a number of implementation issues, suggesting potential hurdles and ways to overcome them.

Policy Developments

In August 2004 President Bush issued a set of executive orders to reform the intelligence community.[4] These orders provide for a national framework that allows strategic planning to draw on all the tools available across different agencies and enables information sharing across the federal government as well as with state and local governments.

Key information-sharing concepts embraced in the executive orders include an agency mandate to revise classification and document management policies to emphasize the "write-to-share" principle (rather than maintaining ownership and control of information in the originating agency) and to use "tear lines" to make the maximum amount of information available by separating out classified elements from the beginning. In addition, the executive orders call for the immediate development of government-wide guidelines and procedures on the sharing of terrorism information that would both empower and constrain government officials as they protect privacy and the security of information.[5] All of these are steps that were recommended and enthusiastically supported by the Task Force.

To establish this information-sharing environment, the executive orders created an Information Systems Council, chaired by a designee of the director of the Office of Management and Budget, "to plan for and oversee the establishment of an interoperable terrorism information sharing environment to facilitate automated sharing of terrorism information among appropriate agencies."[6] As envisaged, the plan must specify the resources, functions, and changes that are needed, as well as the timeline and responsibilities for implementation. Furthermore, the executive orders call for near-term steps, such as the creation of electronic directories, that would link analysts in different agencies and establish which agencies have information on particular subjects of interest.

In developing this plan, one of the great challenges will be designing policies to protect privacy and other civil liberties while expanding the information that the government examines and shares. For example, our Task Force has called for clear guidelines on collection of information on U.S. citizens or permanent residents that would require documentation of relevance, by program or activity, to the homeland security mission. We have also urged expansion of audit and oversight to enhance public confidence. It is critical that these policies be developed in the right sequence with the development of the technology. Otherwise, the public and its representatives in Congress and elsewhere will see concerns where they may not lie because they will not understand the constraints and oversight the government is imposing on itself in the interest of preserving traditional values of individual rights.

Four months after the president issued his executive orders, Congress passed the Intelligence Reform and Terrorism Prevention Act of 2004, which further adopted the framework of the Task Force recommendations for a trusted information-sharing environment. The act, among other things, establishes a Senate-confirmed Director of National Intelligence with extensive authority, a Privacy and Civil Liberties Oversight Board, and a National Counterterrorism Center (integrating the previously created Terrorism Threat Integration Center). It also calls for the creation of a decentralized, distributed, and coordinated information-sharing environment with built-in safeguards for civil liberties.[7] In particular, the act specifies that this environment should

—connect existing systems, where appropriate; provide no single point of failure; and allow users to share information among agencies, between levels of government, and, as appropriate, with the private sector;

—ensure direct and continuous online access to information;

—facilitate the availability of information in a form and manner that enables its use in analysis, investigations, and operations;

—build upon existing systems capabilities currently in use across the government;

—employ an information access management approach that controls access to data rather than just systems and networks, without sacrificing security;

—facilitate the sharing of information at and across all levels of security;

—provide directory services, or the functional equivalent, for locating people and information;

—incorporate protections for individuals' privacy and civil liberties; and

—incorporate strong mechanisms to enhance accountability and facilitate oversight, including audits, authentication, and access controls.

Now the stage is set for real and significant reform of our intelligence environment. These legal developments require a shift from planning for change to making change happen. In this new stage, it is essential to implement the letter and spirit of the executive orders and the Intelligence Reform Act in a manner that truly enhances security while simultaneously preserving privacy and secrecy. In what follows, we discuss some key implementation issues.

Key Features and Implementation Issues

Among the many new requirements, six key features can be highlighted in the suggested intelligence-sharing network environment. Each of these presents specific opportunities (and difficulties) when it comes to implementation.

A Distributed System

Perhaps the most essential (or foundational) characteristic of the network is that it needs to be widely dispersed—across agencies, jurisdictions, and geography. Currently, there exist some fourteen intelligence components in the federal government and in federal intelligence and security agencies. In addition, there are 17,784 state and local law enforcement agencies, 30,020 fire departments, 5,801 hospitals, and millions of other first responders on the frontlines of homeland security.[8] Each of these has access to dispersed bits of data, pieces of a puzzle that, if put together, could save lives. Add the hundreds of thousands of private entities that have similar access to data or information, and the sheer volume and spread of information become evident.

Clearly, such a dispersed pattern of information holding requires a similarly dispersed system for information sharing. Successful counterterrorism cannot rely on a centralized mainframe in Washington or a hub-and-spoke approach. It must instead be "empowered at the edges" and based on a model that includes multiple ways to access information and a system of authorization for appropriate access to information.

Several steps are necessary to implement this goal of a dispersed network. First, it is critical to begin by *leveraging existing networks*. Indeed, Executive Order 13356 (section 5[b]) envisions such a system of systems or "interoperable environment." This approach reduces the need for wide-scale reengineering or the creation of a megasystem. It is not only quicker but also potentially cheaper. By ensuring a certain level of *basic interoperability* between existing networks, we can begin the process of intelligence-sharing across a distributed network almost immediately.

For this network capability to be truly dispersed, a further important criterion must be fulfilled: it must be designed (and interconnected) in a manner that facilitates *horizontal collaboration*. Intelligence must not be shared or accessed through a central hub, which could pose a potential roadblock, but rather through tools (for example, directories, keyword searching, and publish-and-subscribe software) that are available to all relevant and appropriate system participants. Needless to say, to preserve the integrity of data, these tools must also include rules for when and how an actor can access information. As discussed below, the Task Force has recommended that these rules incorporate a requirement for an auditable documentation of a query, as well as an appeal process in case a query is rejected or deemed unnecessary.

In addition to these steps, further consideration must be given to *including nonfederal players* in the distributed intelligence-sharing network. It has become increasingly clear that essential bits of information often reside at the edges of the system—on the frontlines, for example, of law enforcement. It is, therefore, critical that the network include a number of actors beyond the federal government. These include state and local governments, of course, but also private sector entities and foreign governments, both of which may have important information. Indeed, in its second report, the Task Force devotes significant effort to detailing the potential importance of data residing with the private sector. Developing a system (including well-defined access and privacy rules) for accessing some of these data is a critical component in implementing any nationwide intelligence-sharing network.

A Trusted System

A distributed system is critical to effective intelligence-sharing. But if that system is not trusted, both by the public and by those expected to populate it with data, then its effectiveness will remain limited. Government agencies must trust that information will be handled in a manner that does not jeopardize national security. Equally important, the public must be confident that personal information will not be misused and that civil liberties will not be abused. Without such trust the system is likely to encounter significant opposition, both at its inception and throughout its existence. We believe that it is possible to develop an information-sharing environment that enhances our security and civil liberties protections at the same time. However, to do this, the Task Force has strongly argued that the implementation of mechanisms to enhance trust must be built into the system from the very start of the network environment, during its design phase. Fortunately, several mechanisms exist to promote such trust.

To begin with, and as specified in Executive Order 13356 (section 4), the network must be built on *clear and systemwide guidelines* for the collection, handling, distribution, retention, and accuracy of information. Among other things, these guidelines, which should be released to the public, will ensure that personally identifiable information about those who are not suspected terrorists is used responsibly; they will provide clarity about what is and is not permitted, thus reducing the reluctance of officials to share information; and they will ensure that only officials with certain authorization can access particular pieces of information, thus helping to preserve national security. These guidelines should be driven by policy judgments made by politically accountable officials; they are not primarily a statement of legal requirements.

One type of guideline that needs special attention is a requirement for *predicate-based searching*. Predicate-based searching would require officials to demonstrate a clear terrorism link in order to access personal or private information on the network. Predicate requirements can be more or less rigorous depending on the type of information sought and the user involved. In some cases, for example, reference to a particular investigation or program will provide the necessary predicate; in other cases—such as when the user seeks sensitive private sector information about a U.S. citizen or resident—a more detailed explanation of need, or even confirmation of approval, may be necessary. Overall, whenever personally identifiable information about U.S. citizens or permanent residents is involved, the system should be designed to favor predicate-based searching over more general pattern-based or probabilistic queries. These latter types of queries contain greater potential for misuse of personal data not relevant to a terrorism investigation—and thus hold greater potential for eroding trust in the system.

Guidelines are also important in the context of watch lists. It is essential that clear rules be developed regarding, among other things, how parties get on or off lists and what recourse or appeal parties have if they feel unjustly included on a list.

Developing alternatives to traditional classification systems is another area of challenge in writing guidelines. In sharing domestic and foreign intelligence information, we need to find new boundaries that protect the security of critical information as well as civil liberties.

All these guidelines, of course, will have little relevance unless they are respected. Therefore, a trusted network environment must also *incorporate a range of tools, technical and nontechnical, to ensure that guidelines are properly followed*. Such tools can include, for example, automatic tracking and auditing technologies, which allow for oversight and accountability in the way information is used. In addition, strong authentication, encryption, and permissioning technologies are critical. These technical tools should be supplemented by some form of human oversight that monitors the system on a regular basis. The human component can also include training, spot and periodic reviews, and systematic and proactive communication of the system's values to those who use it. Clear lines of responsibility—leading up to Congress—should be specified.

Finally, internal governmental safeguards to promote trust should be enhanced with *external advice*. While the executive order and the Intelligence Reform Act both envision the creation of civil liberties boards (the Executive Branch Civil Liberties Board and the Privacy and Civil Liberties Oversight Board, respectively), these boards should be supplemented by outside and independent participation in policy development and monitoring. Such external input will be critical in winning public trust.

Designed around a "Need to Share"

Our existing intelligence-sharing paradigm is based on a "need to know" principle: it assumes that information secrecy is of paramount concern and restricts access to a select group of individuals and agencies. This principle might have been effective when the enemy was clear and well defined. But in our current situation, where threats are dispersed, and where clues to attacks may lie in the most apparently innocuous pieces of information, a broader range of law enforcement and intelligence agencies must have access to intelligence.

This new paradigm, designed around a "need to share" principle, does not require us to compromise on national security. One of the key challenges confronting implementation will be the need to *balance the need to share with the need to preserve information secrecy*. Each of these goals is critical to national security, and they are not mutually exclusive.

To achieve the necessary balance, the Task Force has recommended adoption of a *risk management strategy* that would help protect sources, methods, and other highly sensitive information. It is necessary to accept, however, that one hundred percent security is unattainable; a "risk assurance" approach that requires this level of security would place an unacceptable burden on the network. Instead, the Task Force has proposed a *layered approach* to security, which promotes sharing while minimizing risk through appropriate use of technology, policy, and oversight.

An essential tool in this risk management strategy is the use of *tear lines* (as envisioned by Executive Order 13356, sections 3[a] and 3[b]). Such tear lines can be used to separate information from its underlying sources and methods; this makes intelligence easier to disseminate. For example, a report could include a paragraph explaining sourcing in a classified section under the text that is written to share, separating information on sourcing from the information itself; the contents of the report above the tear line could be shared with lower classification levels.

Technology, too, plays a critical role in promoting tear lines (as well as in other aspects of information sharing). Technology can be used, for example, to electronically separate the classified portions of a report ("below the tear line") from those that are unclassified ("above the tear line"). In addition, technology can be used to "scrub" data, removing classified information (for example, a source's name) from a report before it is distributed. The various electronic authentication and authorization technologies described above will, of course, also be critical in implementing the risk management strategy proposed by the Task Force. And other innovative technologies (analogous to those used on eBay and Amazon.com) can help rate the reliability of sources and point to other interested parties and related threats.

Finally, in addition to using the right technology, certain *cultural changes* must be implemented to enhance sharing. Such changes can be facilitated by the use of personnel incentives (as suggested in Executive Order 13356, section 3[e]), training, and other measures to counteract the cultural and bureaucratic incentives to control information. On a need-to-share network, information belongs to everyone involved in the fight against terrorism; we need to minimize information silos and the compartmentalization of information.

A Redundant, Automated System

The fourth characteristic of the new intelligence environment is redundant and highly automated network capability. Both redundancy (which means that there are multiple electronic pathways to the same information) and automation increase the likelihood that intelligence will be analyzed in a meaningful way and acted upon. These features, which leverage the technical nature of the network, supplement the human capabilities in our current intelligence-sharing system.

Two steps are necessary to implement automation and redundancy. First, the network environment must be designed to *facilitate both horizontal sharing* (among players at the edges) and *vertical sharing* (from the edges to the center, up the chain of command, and back). Currently, sharing, to the extent that it exists, is done primarily from the center down to the edge—the flow of traffic is one way. But a truly dispersed network capability would allow information to flow in many directions. This will increase the chance that analysts will stumble across valuable pieces of information.

Second, certain *electronic tools can be used to achieve this wider flow of information*. These should be built into the design of the network environment and implemented from the start. For example, browsable directories of information can allow users to discover experts, services, and information of which they might not otherwise be aware. Electronic pointers and indexes can fulfill the same function of helping multiple users at various levels "connect the dots."

Increase Knowledge and Support Analysis

In addition to encouraging sharing, the intelligence environment must facilitate analysis and increase knowledge. It must help analysts deal with the inevitable data overload that will afflict the network; it must allow them to separate signal from noise. The Task Force has suggested several steps that can be taken toward this goal.

First, the two features outlined above (automation and redundancy) can also be useful here. *Redundancy*—having *multiple analysts* deal with the same piece of information—increases the chances that that information will be analyzed in a meaningful way. Likewise, *automatic tools* and other applications to circulate information more widely can increase knowledge and support analysis. For example, a "publish and subscribe model," which pushes information to various edges, can be instrumental in making sure that the right pair of eyes sees the relevant information.

In addition, knowledge and analysis can be facilitated through *cultural and organizational changes*. Systematic changes to encourage collaboration, for example, can play a crucial role; these could include the creation of ad hoc analytical teams to share information and coordinate efforts. More formal arrangements, such as assigning certain personnel to a strategic coordinating role, can also be useful.

Continuous Development and Change

Finally, an effective intelligence environment must be built on processes that are flexible enough to permit continuous change and evolution. Confronted with a fast-moving, asymmetric, and evolving threat, a static, unchanging system would be wholly inadequate. The system must learn from experience; it must undergo regular upgrades of both technology and policy, using past performance as a guide. To facilitate this process of evolution, the Task Force has suggested two important steps that should be built into the environment.

The first is *designating a senior official* in the Executive Office of the President to have full-time responsibility for developing and maintaining the vitality of the environment. This official would watch the system, evaluate its performance, and evaluate the environment within which that system operates. He or she could then suggest upgrades or modifications based on changes in the nature (or level) of threats confronting the nation, as well as in the technical, policy, or other realms. The Intelligence Reform Act establishes a "program manager" who may implement the above tasks if given sufficient authority.

To adequately perform these tasks, this official will require *clearly defined metrics* by which to measure the system's progress. These metrics, which would use objective performance measures, could help quantify how the system is producing and sharing information, as well as its overall effectiveness in countering threats. They would allow the designated official (or any other individual or agency) to gauge the nation's progress toward better information sharing and, ultimately, to assess the state of our national security.

Conclusion

The implementation of a trusted information-sharing network environment presents serious challenges. Achieving it will take years of dedicated and focused work. The policymaking, cultural, and technical changes are substantial. Nonetheless, the stakes are sufficiently high, and we believe that we are finally on the path toward real reform. The decisions we make now regarding the technical, legal, and cultural architecture of the information-sharing environment will have ramifications for generations to come. That is why it is so essential to consider not only what we want out of the network, but just as important, the ways in which we want to achieve those features. This paper represents an initial contribution to that discussion.

Notes

1. Task Force on National Security in the Information Age, *Creating a Trusted Information Network for Homeland Security* (www.markle.org/download-able_assets/nstf_report2_full_report.pdf [December 2003]).

2. See Markle Foundation Task Force, *Protecting America's Freedom in the Information Age* (www.markle.org/downloadable_assets/nstf_full.pdf [October 2002]). Additional information on the Task Force can be found at www.markle.org.

3. P.L. 108-458.

4. See Executive Order 13353, *Establishing the President's Board on Safeguarding Americans' Civil Liberties*; Executive Order 13354, *National Counterterrorism Center*; Executive Order 13355, *Strengthened Management of the Intelligence Community*; and Executive Order 13356, *Strengthening the Sharing of Terrorism Information to Protect Americans*, all issued August 27, 2004 (www.whitehouse.gov/news/orders/ [October 2005]).

5. Terrorism information is defined as that which "relates to" foreign, international, or transnational terrorist groups or individuals, providing an important limit on the kinds of information that can be part of a shared information environment and thereby enhancing public trust that information unrelated to terrorism will not be brought into the system. Executive Order 13356, *Strengthening the Sharing of Terrorism Information*.

6. Ibid.

7. P.L. 108-458, sec. 1016. The Senate version of the bill that included the attributes passed by Congress explicitly referenced the recommendations of the Markle Foundation Task Force.

8. For sources for these figures, see Task Force on National Security, *Creating a Trusted Information Network*, p. 11.

5

Security and Liberty: How Technology Can Bridge the Divide

GILMAN LOUIE AND GAYLE VON ECKARTSBERG

O n October 25, 2001, one day before the PATRIOT Act rolled out, Attorney General John Ashcroft issued a warning to terrorists within our borders: "It has and will be the policy of this Department to use . . . aggressive arrest and detention tactics in the war on terror . . . we will use all our weapons within the law and under the Constitution to protect life and enhance security for America."[1]

While the vast majority of Americans agree that action must be taken to secure our country, many fear that new legal authorities might come at the expense of our civil liberties. Central to this debate is whether or not the U.S. government should be allowed to use data on U.S. citizens to help ferret out terrorists. Fifty years ago cold war fears fueled probes into personal lives and led to blacklisting and McCarthyism, all in the name of national security. What is more disconcerting today is the sheer volume of information on our citizens that is held in private and government hands—information that is increasingly available because of powerful technologies that have emerged in the last five years.

Yet information and technology are critical advantages that the United States has in the war on terror. And in the great debate between national security and civil liberties, we should not automatically assume there must be a trade-off. As a matter of principle, the question of whether to protect civil liberties or ensure our national security is a false choice. And in the most practical sense, as argued by the Markle Foundation Task Force on the Information Age, the technology available today makes it possible to do both.[2] In this chapter we will focus on some of the new tools that make it possible to preserve our security and our liberty at the same time

The World is a Dangerous Place

Today, it seems trite to say that we live in a dangerous world, but before September 11 this was not so obvious. The post–cold war world was a blurred and changing environment marked by the quiet proliferation of weapons of mass destruction, the emergence of rogue states, and the rise of nonstate actors. Throughout the 1990s U.S. intelligence organizations came to understand that in addressing these challenges, our greatest strength lay not in big power politics but in our technological superiority and vast information resources.

In the post–September 11 world, there is little need to explain the dangers Americans face, and the urgency our government feels to act. The threat of terrorism against our homeland and national interests can come from anywhere, overseas as well as within our own borders. Decentralized and asymmetrical in nature, terrorist organizations act in ways that require a response in hours instead of weeks or months. The threats are as varied as the imaginations of those who aim to harm us. They may use weapons of mass destruction—nuclear, biological, or chemical—or they may use simple tools to terrorize and disrupt. Their activities may involve years of meticulous planning. Or, as the nation learned in October 2002, it may only require two Americans with a sniper rifle, a laptop computer, and a car to paralyze the nation's capital.

A New Kind of Response

To deal with such threats effectively, the government must have the ability to collect, share, and act on information more rapidly than ever before. Unpredictable, adaptable, and disciplined, terrorists seek to disrupt our way of life as much as to destroy it. To counter the threat, the United States must turn the tables and disrupt their operations. We must be able to observe, orient, decide, and act faster than our adversaries. To paraphrase the concept developed by Colonel John R. Boyd in the 1970s, our ability to prevail depends on having superior information and intelligence in the hands of our policymakers, analysts, war fighters, and first responders. We must rapidly process and deploy information and decentralize decisionmaking in order to enhance the speed, flexibility, and responsiveness of our intelligence machinery and workforce. Agility enables us to move quickly and decisively; it enables us to be more unpredictable than our opponents.

The institutional framework that served the nation well through the cold war is now changing to address this new mission. Created in 1947, the Central Intelligence Agency was charged with coordinating the nation's intelligence activities but prohibited from having any "police, subpoena, law enforcement powers, or internal security functions." Law enforcement, as well as counterintelligence, was part of the brief of the Federal Bureau of Investigation. In the following decades, both agencies built internal systems, processes, and cultures to address known enemies with known capabilities employing known tactics. Together, they created a complex legacy of separate databases, multiple levels of classification, compartmentalization, and diverse business practices and legal frameworks that persist today.

Some of the challenges facing our intelligence system became obvious after the September 11 attacks. The attacks revealed important gaps in the way the government analyzes and shares information. They also highlighted another underlying problem: that our ability to collect information may well outstrip our ability to effectively use it. The Congressional Joint Inquiry that examined the work of the intelligence community found no information that, if fully considered, would have provided specific, advance warning of the details of those attacks. But it also observed that "within the huge volume of intelligence reporting that was available prior to September 11, there were various threads and pieces of information that, at least in retrospect, are both relevant and significant. The degree to which the Community was or was not able to build on that information to discern the bigger picture successfully is a critical part of the context for the September 11 attacks."[3]

Leveraging What We Already Know

Today, the intelligence community must address two fundamental challenges: first, how to transform the vast amount of information available into knowledge; second, how to close the gaps in sharing and analysis of information, not only within intelligence organizations but across federal, state, and local governments, reaching cops on the beat and analysts at every level.

The government requires broader and more effective access to information but not necessarily more information. Much of the information it needs is already being collected and held by local law enforcement officials or by other federal, state, and local agencies. In addition, the private sector holds a valuable trove of information on U.S. citizens and others within our borders. In fact, in the past ten years, the private sector has become the dominant player in the capture, organization, and use of personal data—data that at face value seem trivial. But credit card transactions, flight information, rental car records, and phone logs all have the potential to help identify the next terrorist or help alert us to a potential attack.

Sophisticated processing of vast quantities of data and the profiling of individuals by the thousands are everyday activities in today's marketplace. When we use a credit card, shop at any major retailer, or purchase an airline ticket, our transactions are recorded. Companies study these data to identify inside traders, to target their marketing more effectively, or to decide when to replenish their shelves. It is increasingly possible to do this analysis in real time and use it to predict future behavior. An entire industry specializes in data aggregation—gathering and selling such data to commercial companies, the government, and individuals.

The government's ability to access and utilize this information is essential to the war on terror. Quite simply, without data derived from credit card purchases, travel reservations, financial transactions, telecommunications activity, and the Internet, it would be extremely difficult to locate, monitor, and prevent known terrorists and their associates from carrying out their activities. Historically, government agencies had access to such information on a case-by-case basis. Upon request, companies would either volunteer the information or respond to a subpoena demand. In other instances government offices might require specific collection and reporting, as in the case of financial services. But, by and large, commercial information remained in commercial hands.

Now several factors have opened the way for government to exploit the full range of available public and private data. With the advent of commercial data aggregators, as well as the availability of customized datasets, the increasing affordability of computation and data storage, and exponential improvements in computer performance, many agencies are seeking to exploit such data more systematically. The goal of these initiatives, which span federal, state, and local law enforcement agencies, is to find the "bad guys" by combining information on commercial transactions with government-held data and then "mining" the results to identify potential threats.

Refining a Blunt Instrument

The promise of data mining in the war on terror seems straightforward: with today's technology we can rapidly process the data we have and quickly identify profiles of potential threats. The challenge is that when looking for patterns in vast amounts of information, one can see any correlation one is inclined to see.[4] For this reason many privacy advocates argue that giving the government unprecedented access to large repositories of personally identifiable information could result in profiling, via the use of computer algorithms. Individuals could be tagged as potential threats based on their race, country of origin, socioeconomic background, or personal buying habits. But flagging people as risks based on any combination of criteria—name, origin, age, travel itinerary, or even the books they've read—can result in false positives, as well as identify potential threats. Consequently, this approach could lead the government to infringe on innocent citizens' rights—or even arbitrarily tag them for life.

Contributing to this concern is the lack of transparency and accountability surrounding the algorithms and methodologies used by large data crunching systems to score potential threats. Some observers fear that decisions that affect our civil liberties are being made by computers, using unproven profiling rules, not by human beings operating with transparency and accountability.

Finally, civil libertarians worry about mission creep. While data may be collected today for use in the war on terror, there are no guarantees that they will not be used by the government for other purposes down the road. The question, then, is how does the government leverage this wealth of information to protect our nation without compromising our civil liberties?

Alternative Approaches to Data Mining

One answer lies in choosing the right approach to data mining. Current tools for data mining employ two distinct methods: data profiling and data analysis. The first is a form of pattern analysis; the second relies on link analysis.

The profiling approach is akin to personal credit scoring. Most Americans have encountered credit evaluation systems and know how unreliable and arbitrary they can be. With this approach computers may flag individuals as permanent credit risks based on something that took place in their past. As a result, it may become difficult, or even impossible, for a person thus categorized to ever buy a home or a car.

This technique is a shortcut—a simplistic approach to a complex problem. It could be applied to the war on terror by establishing a system that, for example, does the first sort on people with Middle Eastern names, the second sort on naturalized U.S. citizens, the third sort on people who have traveled overseas in the past six months, and so on. Despite the use of sophisticated technology, this approach is not too different from the racial profiling that occurred during World War II, when Japanese Americans, such as California representative Mike Honda, were forced to live in internment camps simply because of ancestry. Computerizing the process of profiling makes it more efficient but neither better nor right.

If this approach is adopted, the government should be held to much higher standards of accuracy, fairness, and transparency than credit rating institutions. Financial institutions are not required to disclose how they develop your credit rating or which criteria or data they employ. Taking security risks into consideration, government agencies should be open about their methods. They should also take the time to do more thoughtful analysis or invest in the technology systems needed to do the job.

Alternatively, officials may choose the data analysis approach. In contrast to data profiling, this method puts the emphasis on tracing links from known individuals or known data. It puts human analysis at key junctures of the process and focuses on leveraging information the government already has to drill in on the next possible threat.

Data analysis can be a powerful investigative tool for tasks such as matching individuals against watch lists of known terrorists. It begins with a lead or a predicate—a real piece of intelligence—and then follows the people, the money, and the information flow, link by link. But even this analysis only provides a starting point for further human analysis. Someone must still follow up—by interviewing potential suspects, for example—to find out if the purported connection is real or coincidental—a matter, say, of two people having the same birthday or same address.

Effective analysis cannot rely on computer systems and algorithms alone. Those are just tools to reduce the noise in the set of possibilities. By starting with a lead or known fact regarding a threat, human analysts can use data mining as a highly targeted instrument to construct investigations and threat analysis and to identify actionable information. This is the advantage of using the right technology, targeting the known facts, and giving analysts more time to focus on valuable information.

New Technologies and Tools

In addition to alternative approaches to data mining, government agencies also have access to a growing range of tools for protecting personal information. The trade-off between civil liberties and national security is a false choice. Simplistic approaches—fishing expeditions into personal data or creating a massive government database of private information—overlook the tools now being developed in the private sector. New data exploitation capabilities are being created every day, fueled by innovations in supercomputer architecture and in data collection and management. If used appropriately, these technologies can provide an alternative to blunter instruments and enable government to use private information while protecting civil liberties. Below we identify five approaches that can help make the trade-off between national security and civil liberties unnecessary.[5]

SELECTIVE REVELATION. Rather than make all information freely available, this approach selectively reveals information to the analyst according to business rules and criteria. For example, key criteria may include, for example, who the analyst is and the status of the investigation. Authorization rules then determine what information is revealed to whom and at what time. To retrieve additional personal information, a higher level of authorization might be required, based on an independent evaluation (by a court, for example) of the evidence that the analyst is actually "on to" something suspicious.[6]

ANONYMIZING. By employing techniques such as hashing, masking, and blind matching, which effectively render information anonymous, analysts can search for potentially dangerous patterns without requiring access to large amounts of personal data. These anonymizing capabilities can be secured by encryption and be synchronized and updated by authorized agencies. In addition, policy management and audit tools can be built into such systems.

Commercial enterprises can also use anonymizing technologies to screen their data holdings for specific activities or watch list matches and report only the necessary information to the appropriate agency when there are specific correlations. This approach keeps personal data at the commercial enterprise, helping to prevent the government from amassing large databases of private transactional information.

DIRECTORIES AND DECENTRALIZED POINTERS. If building a mega-database controlled by government is not the answer, then how can government effectively use information that is widely distributed across commercial enterprises and across the government itself? A system of directories that point the way to organizations with particular information stores is one means for the government to access what it needs, without taking ownership of sensitive data.

AUDIT AND ACCOUNTABILITY. Audit and accountability technologies answer the question, "Who used the system to retrieve what data and when?" These tools can help protect personal information by making it exceedingly difficult for unauthorized personnel to access and use data systems without being detected. They can also help create the trust and accountability necessary for government organizations to share information effectively as well as promote public trust by enabling individuals filing Freedom of Information Act requests to learn how personally identifying information has been used.

Commercial technologies, such as the electronic data processing auditing techniques used by accounting firms to audit financial data and control systems, can easily be applied to government data mining systems. These highly developed technologies provide built-in recording capabilities that track how information is accessed, used, retained, and shared. They also make it possible to correct erroneous or out-of-date data points and to ensure that the same corrections are made to every piece of analysis using that information. In the absence of such tools, inaccurate database references could easily dog individuals for life.

Similarly, agencies handling sensitive information can use digital rights management technologies to limit the use of data to a particular purpose for a particular period of time. These tools allow the original data owner to enforce its security policies, even after data have been distributed to different users. For example, the original owner can cause a file to expire at any time. This means that all copies will be automatically deleted wherever they are located, whether on users' desktops or on servers and other systems.

ENTITY RESOLUTION. Entity resolution involves such questions as, "Is this the person we are looking for? Are this address and this geolocation or name of a building one and the same? Who owns this phone number?" Names alone pose multiple problems: they have multiple transliterations, diminutives, and alternative spellings, which make them inadequate as definitive indicators of identity.

Without entity resolution, watch lists can be cumbersome and not very effective, leading to false positives as well as false negatives. However, a range of commercial entities, including credit companies, data aggregators, and fraud analysis services, can contribute their expertise in this key area. These enterprises have a similar need for entity resolution and actively develop and source technologies to support analytical services.

Conclusion

Several high-profile government efforts to leverage private personal information in the war on terror have been pulled up short. Total Information Awareness, the Multistate Anti-Terrorism Information Exchange program, and the Computer-Assisted Passenger Prescreening System have all run headlong into intense public resistance—in part due to concerns that these programs would infringe on civil liberties.[7]

Today, fearing a public and congressional backlash, many agencies are reluctant to acquire advanced analytic capabilities and tackle the privacy issues necessary to clear the way for appropriate use of transactional information. Yet doing nothing is not the answer. The risk of this approach is that if another major incident were to occur, the government might be forced to act hastily, without taking the time to develop policies that adequately account for citizens' rights. The end result could well be a significant compromise of civil liberties.

Tools exist today to enable the government to use private information while minimizing the impact on individuals' privacy and civil liberties. What is needed as a first step is for the government to articulate and implement clear policies. There must be guidelines to clarify what private information the government will access and how it will be used. Such guidelines should also establish the privacy standards that will govern the use of technologies to access this information.

The second critical step is to put someone in charge. To this end, the Intelligence Reform and Terrorism Prevention Act of 2004, consistent with the findings of the September 11 Commission, provides for creation of a Privacy and Civil Liberties Oversight Board and establishment of a Civil Liberties Protection Officer within the Office of the Director of National Intelligence.[8] In creating these positions to oversee government data access and sharing programs, Congress has demanded accountability and responsibility for the protection of civil liberties.

Finally, there must be a fundamental change in the way the government does business. This new approach must bridge gaps in the analysis and sharing of information across agencies and federal, state, and local governments. And it must leverage the vast information resources the government already has.

Our government today is engaged in a tremendous transformation, the likes of which has not been seen since World War II. Dedicated professionals across government are remaking their organizations and taking bold steps to find new ways to help make our nation more secure. They have an urgent mission, and they have the support of the American people. Recent legislation and leadership from the White House have opened the way for significant progress. What is needed is transparency in our policies and action from our leadership that will assure us that our civil liberties as well as our nation's security will be protected.

Notes

1. Attorney General John Ashcroft, "Prepared Remarks for the U.S. Mayors Conference," October 25, 2001.

2. Task Force on National Security in the Information Age, *Creating a Trusted Information Network for Homeland Security* (www.markle.org/download-able_assets/nstf_report2_full_report.pdf [December 2003]).

3. Senate Select Committee on Intelligence and House Permanent Select Committee on Intelligence, *Joint Inquiry into Intelligence Community Activities before and after the Terrorist Attacks of September 11, 2001*, H. Rept. 107-792 and S. Rept. 107-351, 107 Cong. 2 sess. (Government Printing Office, December 2002), p. 7.

4. Markle Foundation Task Force, "Part 2: Working Group Analyses," *Protecting America's Freedom in the Information Age* (www.markle.org/downloadable_assets/nstf_full.pdf [October 2002]).

5. For more details, see Task Force on National Security, *Creating a Trusted Information Network*.

6. See House of Representatives, Committee on Government Reform, "Statement by Dr. Tony Tether, Director, Defense Advanced Research Projects Agency, before the Subcommittee on Technology, Information Policy, Intergovernmental Relations and the Census," May 6, 2003 (www.fas.org/irp/congress/2003_hr/050603tether.html [December 2005]).

7. For brief discussions of these programs, see the "Liberty and Security Timeline" at the end of this volume.

8. P.L. 108-458.

6

Policies and Procedures for Protecting Security and Liberty

BRUCE BERKOWITZ

The information revolution is creating new threats to U.S. security. Thanks to modern technology—the ubiquitous Internet, affordable satellite phones, cheap computers, and easy-to-use strong encryption— terrorist organizations can deploy covertly within our borders and strike with little or no warning.

True, the United States has always faced a potential threat of sabotage and terrorist attack. Sabotage against American cities and other nonmilitary targets occurred in the War of 1812, the Civil War, World War I, and World War II. Recently, former Soviet military officials have claimed that the Soviet Union would have conducted sabotage against the United States—possibly with nuclear weapons—if the cold war had turned into open conflict. But today technology has transformed what had been a strategy of desperation into a preferred course of action. In the past, terrorists, guerrillas, and special forces operated as small fragmented units with limited firepower, unable to communicate with their leaders for weeks or months. Now information technology makes it possible to weave such units into a network of integrated cells able to communicate reliably with counterparts around the world.

Such networks can remain under cover and strike at a time and place of their choosing to maximize both destruction and fear. To make matters even worse, such networks today could potentially use nuclear, radiological, chemical, or biological weapons. Indeed, network warfare tactics solve the problem that discouraged attackers from using such weapons in the past. Clandestine networks often leave no calling cards, and that makes retaliation—and thus deterrence—harder.

This kind of warfare is so potent that armies as well as terrorists are adopting the basic tactics of stealth and networked command systems. In future wars U.S. officials will need to consider the real possibility that an adversary will use this approach. Because the United States is overwhelmingly powerful in conventional military forces, our potential adversaries will seek an alternative. Network warfare provides one. The necessary technology is widely available commercially, and so almost any nation with a grudge and a determination to deploy a global, disciplined, networked fighting force can do so.

Taken together, these developments are rendering obsolete the traditional idea of the battlefield bounded by a front line separating two armies. To use modern parlance, in the future the "domestic battlespace"—the land, sea, and air terrain of the American homeland—could easily become a major theater of war.

This points to the problem and the mission: how does the United States search for foreign adversaries within its own borders without sacrificing its citizens' civil liberties? Technology can help resolve this dilemma but not by itself. If we hope to carry out this new mission successfully, then we must also create a new kind of intelligence organization and develop new policies and procedures that create a safe zone for collecting and sharing information. These policies and procedures are the focus of this chapter.

Intelligence Requirements for Homeland Defense

Detecting hostile networks within the U.S. homeland presents an intelligence problem that is fundamentally different from the one the United States faced during the cold war. Then the threat was well known, and the potential channels of attack were limited. The most likely scenarios involved a Warsaw Pact conventional strike on Western Europe across the inter-German border or a Soviet nuclear attack on the United States from approximately 1,600 fixed missile silos and several submarines staged off our coast and in the Arctic Ocean.

The intelligence challenge consisted of focusing on this known threat and detecting the signs of an imminent attack. Those signals were well understood because the Soviet armed forces (like other traditional armies) relied on carefully developed, formal war plans. Because of the complexity of these plans, the Soviet armed forces rehearsed them regularly in exercises. We watched the exercises, and that told us what to look for—the "signatures" of an attack.

To detect these signatures, U.S. intelligence relied on a relatively small number of specialized data sets—signals intelligence (SIGINT), satellite imagery (IMINT), and information from a few highly placed human sources (HUMINT). Admittedly, to flesh out the threat so that we could plan our own forces, U.S. analysts used many other sources, but these specialized ones were the most important, especially for tactical warning. Because these sources were sensitive and vulnerable to countermeasures, they were highly classified.

Such "compartmentation" constrained the number of analysts who could work on the problem. But that did not matter much because the kind of monitoring that was required could be carried out by a stable population of analysts. Most of these analysts had highly specialized skills. Most of them worked on the same general assignment throughout their careers.

In contrast, adversaries organized as networks are more likely to change tactics or their direction of attack and hide under many guises. As a result, detecting a potential attack from these new threats is no longer a problem of focusing on information from known sources and having a few specialized experts wait for a predictable warning sign. Rather, it is a problem of gathering information from as many sources as possible and sharing it among a diverse set analysts, and hoping that someone will recognize a pattern that tips off an attack.

In the case of September 11, such a pattern might have been, say, data indicating that marginal students from Egypt and Saudi Arabia who had previously traveled to Afghanistan were now visiting the United States and seeking pilot training in large jets. Or it could have indicated that tourists who had traveled to Afghanistan for six months were now seeking entry to the United States. Or it could have been data evoking an earlier episode when al Qaeda terrorists planned to hijack an airliner and fly it into CIA headquarters. Or it could have been a single sensitive communications intercept or public pronouncement that just seemed to "jive" with past experience of al Qaeda attacks.

The point is, officials cannot know ahead of time who will be able to see the pattern. It depends much on each analyst's background, temperament, available data, and context. So the only solution is to put the data in front of as many eyes as practical to improve the odds. The more eyes—and the more varied perspectives—one can bring to bear, the greater the probability that someone will recognize a potential attack before it occurs. As a result, today homeland security poses a whole new set of operational requirements.

Data

Intelligence organizations now need a wider range of data than in the past. In addition to traditional sensitive intelligence data, they need information from local governments, open sources, and possibly even the private sector. Analysts need the freedom to test many hypotheses—however implausible—if they hope to get a full picture of the domestic battlespace. Because the most sophisticated attackers staging covertly within U.S. territory will avoid attention by obeying the law up to the moment that they strike, the potential to collect information about innocent Americans in the process is almost unavoidable.

Communications

Intelligence organizations need to communicate better—both among themselves and with outside sources of information and expertise. This is partly a technical issue, but it is also an organizational issue. The Joint Inquiry investigating the September 11 intelligence failure reported that the CIA did not inform U.S. law enforcement about two al Qaeda members who later entered the United States and took part in the attacks. This specific error received much attention, but its true importance was not that a particular analyst did not do his job. Rather, this failure was a symptom of a larger problem: organizational turf, training, and security procedures create communications barriers among agencies and analysts.

Products and Consumer Support

Intelligence organizations must now deliver their products to a wider variety of consumers. In addition to federal officials and military personnel, intelligence for countering homeland threats must be shared with local police, emergency responders, transportation security workers, and even much of the public, so that they can prepare. This greatly increases the complexity of a communications network, especially if the network must support people with a variety of security clearances—or none.

Information technology can help. But the most useful technology is probably not the kind that gets the most attention. Ultrasophisticated data mining programs that can sort through thousands of records to identify a particular al Qaeda terrorist from a scrap of information work better in Tom Clancy novels than in the real world. Computers that instantly recognize fingerprints and facial patterns are, at least for now, plot devices for television series like *CSI: Crime Scene Investigation.*

The information technology that would most help intelligence agencies deal with homeland security threats today is much more mundane. Generally, the most effective technology does not try to replace intelligence analysts but rather seeks to make the traditional tasks of analysis—collating and matching data, visualizing relationships, and simply exchanging ideas with other analysts and experts—easier and faster. Analysts are a scarce resource, and technology can potentially enable intelligence organizations to use them more effectively.

Many of these technologies are already available. Notable examples include

—virtual networks that enable analysts to share data and insights more effectively, while also protecting classified information;

—secure teleconferencing systems based on personal computers and the Internet that are cheap enough to deploy among hundreds of thousands of users, and enable analysts and users to accomplish in a fifteen-minute conversation what used to require a day-long meeting;

—computer programs that allow analysts to easily rearrange and graphically present databases on suspected terrorists and their networked organizations;

—search engines that allow analysts to find data that they intuitively know are useful and located in a database;

—commercially available, easy-to-use strong encryption that makes it possible for experts in the private sector to communicate with each other and with government officials as securely as intelligence and military organizations have in the past; and

—software packages that make it easier for analysts to move and manipulate information from a variety of sources, so they can deliver a product to a user in whatever form—memos, briefings, e-mail, or web pages—is fastest and most effective.

Policies and Procedures

Yet technology may be much less important than the policies and procedures that will determine whether it will be adopted and how it will be used. The record of U.S. intelligence in adopting these technologies is spotty—despite the fact that experts inside and outside government have proposed using them throughout the past decade, and especially for homeland defense.[1] There are three main obstacles.

—Security policies: the current rigid, inflexible approach to security is probably the single most important obstacle to better information. Intelligence agencies have been so concerned that an adversary might hack into their computer networks or that a lost floppy disk might compromise sensitive data that they have been slow to invest in new information technology. Even worse, many analysts have been conditioned to think that information technology is more of a threat than an asset, so they are not thinking how such technology could make them more effective.

—Acquisition regulation: cumbersome government acquisition procedures are also a factor. Buying and fielding a single software tool can typically take years. Modernizing the information technology architecture of an entire analytic organization can take so long that it becomes, in effect, a never-ending cycle of studies, planning, and revisions.

—Orthodoxy in analytic tradecraft: technology may make it possible for analysts to use data and expertise from almost anywhere, but their training does not teach them how to do so. There are also few incentives for analysts to use these data creatively, and they are usually not given time to do so in any case. Also, the new technology, which allows individual analysts to interact directly with consumers, conflicts with tradition, which says intelligence analysts should be kept at arm's length from consumers, to avoid tainting the analytic product, and speak with a corporate voice.

Until these issues are addressed, intelligence organizations will be unable to use current information technology effectively, let alone more advanced technology that might come along in the future. [2]

Homeland Intelligence and Potential Threats to Civil Liberties

Today's new requirements for intelligence are not only challenging; they seem designed to raise as much controversy, concerns, and questions about civil liberties as possible. Consider the above requirements, and the specific issues they evoke.

Intelligence Agility versus Rule of Law

Homeland defense requires wide access to data, but intelligence and law enforcement agencies operate under significant constraints that narrow their focus. The CIA is prohibited from collecting domestic intelligence. Although it is able to operate within the United States when its targets are foreign, terrorists and covert military forces will, of course, try to hide their foreign links and mix in with the general public. The National Security Agency operates under even more stringent restrictions and is prohibited not only from collecting intelligence within the United States but in most cases on U.S. citizens abroad.

Early Detection versus Presumption of Innocence

Homeland defense often requires collecting information about individuals who have committed no crime. Testing out "what if" hypotheses about potential threats inevitably means collecting and analyzing data about ordinary, innocent Americans. Many of the analyses one would logically want to carry out for homeland security would violate the law—in particular, the Privacy Act.[3] Moreover, the FBI cannot begin an investigation against a person unless he or she is a suspected criminal, or there is specific information that suggests he or she may take part in a crime.

Sharing versus Privacy

Detecting potential attackers requires sharing information. But if law enforcement organizations share information widely, they may violate a person's statutory privacy rights—not to mention compromise a case they hope to prosecute.

Security versus Participation

Intelligence officials are, by statute, required to protect their sources and methods; this is the basis of compartmentalization. Yet it is unreasonable, and legally questionable, to require police, emergency responders, security workers, and local officials to submit to the kinds of background investigations required for a security clearance.

Controversy and Defensiveness

The aforementioned problems would be hard enough if they were just strictly a matter of statutes and regulations. But one also has to consider the atmospherics associated with intelligence collection. It is not as though the typical CIA intelligence officer or FBI special agent is eager to aggressively collect gobs of information to analyze about the general American public. These agencies are all too aware of the controversies and investigations of the 1970s, and the FBI (deservedly or not) carries the added burden of the Red Scares of the 1920s and 1950s.

As a result, each of these organizations has developed a culture that is reluctant to incur the wrath of the public, the media, and political overseers by undertaking anything resembling these episodes again. This culture is embodied in the minds of their staff members. It is even more deeply ingrained in the minds of legal counsels deployed in depth within each organization and determined to keep their agencies out of trouble.

Improving Intelligence and Protecting Civil Liberties— the Current Approach

Since September 11 the main approach to resolving these problems has been to "lower the bar"—that is, reduce the barriers that preclude intelligence and law enforcement agencies from investigating individuals and sharing information. The first example was the joint resolution Congress passed on September 18, 2001.

The resolution gave the president (and, by extension, executive branch agencies) authority "to use all necessary and appropriate force against those nations, organizations, or persons he determines planned, authorized, committed, or aided the terrorist attacks that occurred on September 11, 2001, or harbored such organizations or persons."[4] In effect, under the resolution, if one could link a party to September 11, he or she was fair game.

This was understandable since the attackers had just killed 3,000 people (the toll was thought even higher at the time), and the government was, in effect, declaring war. Even so, some critics believed these authorities were excessive. Some compared them to Abraham Lincoln's suspension of habeas corpus during the Civil War or Franklin Roosevelt's internment of Americans of Japanese descent during World War II.

Yet even the critics missed the more relevant point: Congress granted the president this authority only *after* the attack. It would not have given the authority if the attack had not occurred. So, logically, this kind of special authority is not a solution to *preventing* strikes by potential attackers of the American homeland.

On the other hand, prevention was precisely the goal of the USA PATRIOT Act, which President Bush signed into law on October 26, 2001.[5] It authorized several measures designed to improve the government's ability to detect foreign threats operating in the United States. These included provisions that allowed law enforcement and intelligence organizations to work together more closely and eased the requirements for approving wire taps, surveillance, and subpoenas.

In both cases, however, the strategy was the same: remove restrictions on the activities of intelligence and law enforcement organizations. For many officials this was probably not even a conscious strategy. Rather, it was an unintentional, Darwinist approach. As experience suggested that certain restrictions impeded intelligence agencies from averting terrorist strikes, officials—prodded by the public—searched for ways to make the agencies more effective. When they identified a particular rule or restriction that seemed to be an obstacle, they proposed removing it.

The main restraint on this process was opposition by those officials and public interest groups with a particular interest in civil liberties. They would mobilize the public and in time create a countercoalition that limited the number and type of restrictions that could be removed. Presumably a scandal or controversy—like Iran-Contra or the 1970s disclosures responsible for many of the current controls—will eventually push public opinion back in the other direction, and the political pendulum will swing back toward more restrictions.

This process—cranking back civil liberties as far as prevailing opinion permits—is all part of democratic politics and is inevitable. But it has several drawbacks, and from a purely strategic perspective, it is clearly not the best way to make policy about homeland intelligence. It leads one to settle on least objectionable options rather than seeking best opportunities. It maximizes controversy—which the war on terrorism was bound to create in any case—and undermines public support for a function that is truly needed. And it frames the debate as a contest between national security and civil liberties, pitting one against the other, even though the two interests are not mutually exclusive.

There are other problems with simply "unleashing" intelligence agencies. Obviously, innocent persons who come under their attention will potentially suffer damages to their freedom, reputation, or bank account. Even worse, it could lead to a net loss in our intelligence capabilities rather than a net gain. It is hard to focus existing intelligence organizations on the homeland security threat without compromising the tradecraft that has made these agencies as effective as they are in their current assignments.

Foreign intelligence tradecraft includes, to put it bluntly, lying, bribing, and using various forms of coercion to obtain information. It also includes dealing with foreign intelligence and police agencies that do not meet our own standards for respecting human rights. And it includes using whatever means necessary to break into electronic communications and data. The question is, do we want (hopefully) aggressive organizations with the skills, motivation, and mindset to carry out foreign intelligence operations to operate within the United States, where they will inevitably affect many Americans? In the past, the answer has usually been "no." Since September 11, the answer has been "maybe."

In this connection, there are costs to lowering the threshold for triggering action by the FBI and other law enforcement organizations. When a law enforcement agency collects information, there is a clear implication that officials believe someone committed a crime or is about to commit a crime. Anyone who comes under their scrutiny inevitably is tagged with some stigma. Also, law enforcement agencies can threaten apprehension, detention, and referral for prosecution, so whenever they act, there is the potential cost that their target will lose time, money, and well being.

All of this is exactly why there are constitutional, statutory, and regulatory constraints on both intelligence and law enforcement organizations. If we loosen the constraints, there is an inherently higher risk of violating someone's civil liberties. However, if we do nothing, we leave ourselves blind to homeland-based threats.

An Alternative Approach to Protect Civil Liberties

Current measures to improve our ability to collect information about homeland threats simply *lower the bar* that restrains intelligence and law enforcement organizations from investigating individuals. Another approach is to adopt measures that *limit the potential damage* of such investigations. Such measures would control and mitigate the harms that occur when the government collects data about individuals and errs by taking action against innocents.

Limit the Ability of Information Collectors to Take Action

Ironically, much of an organization's power to act as an early warning system would result from the authorities that it did *not* have. Currently, the CIA, National Security Agency, FBI, and other federal agencies are prohibited from collecting information that private citizens are allowed to collect, because their powers inherently raise concerns about potential abuse. By specifically prohibiting a new intelligence organization from arresting or detaining people and defining the specific conditions under which it could release information to intelligence and law enforcement organizations, it would be possible to give it greater authority to collect information. If an organization's only mission were to inform officials about threats, then the level of these fears would be lower.

Armed with this combination of powers and constraints, the new analytic organizations could provide such assessments as

—detailed threat assessments of privately owned buildings such as football stadiums, transportation terminals, ports, historic landmarks, amusement parks, and other potential targets;

—profiles of purchasers of chemicals and biological specimens that might be used as potential weapons but that are not controlled under law; and

—personal dossiers of foreign nationals who have been admitted to the United States on tourist or business visas but who do not meet the threshold for conducting a criminal investigation.

Protect and Control the Use of Information

The traditional approach to protecting information usually assumed that the information in question was collected by the government and belonged to the government. This information included, for example, classified intelligence, military operational plans, secret weapons, and law enforcement data. The solution to protecting this information was straightforward: unauthorized disclosure was often a crime.

Homeland security is more complicated in large part because the data relevant to homeland defense belong to private companies, local government, and individuals. It is much harder to make disclosure of such data criminal, and the government has limited leverage in controlling and therefore protecting it.

This is an important problem in the design of any "safe zone" because access to data is usually linked to one's willingness and ability to protect them. For homeland security the solution often will not be classification; it will be a civil agreement, such as corporations make with each other to share and protect intellectual property.

Provide Recourse to Subjects of Mistaken Investigation

Homeland intelligence will, if pursued aggressively, sometimes cause harm to innocent individuals. There needs to be a fair way of adjudicating their claims and, at least in some cases, compensating them. Homeland intelligence officials should be provided standards of conduct and, as in the case of police, indemnified if they follow them.

Additional Measures for Oversight and Control

Formal rules and regulations can only take the system so far. In addition, such safeguards and mitigation procedures must be built into the day-to-day operations of intelligence organizations.

TRADECRAFT. Agencies also have to build an ethic of protecting civil liberties into the training and methods of intelligence officers. All members of a homeland intelligence organization must understand that their success in completing their mission depends on their success in striking a balance between pursuing enemies and protecting civil liberties. There are precedents for this kind of professional code, for example, the ethos that case officers have about protecting their human intelligence sources and the way cryptologists learn to minimize exposing U.S. citizens to signals intelligence activity.

CONSENSUS REPRESENTATION FOR OVERSIGHT AND CRITERIA. Homeland security intelligence will continuously require officials to make judgment calls in matters such as whether a person or group is a legitimate target, or whether to pass information and analysis to consumers or other intelligence and law enforcement agencies. In-house counsel can do part of the job. But a standing body representing a range of political viewpoints should develop the guidelines for these counsels.

GATEKEEPERS AND SWITCHMEN. One of the reasons the United States was not more effective in dealing with al Qaeda prior to September 11 was that officials could not decide whether the threat was a law enforcement problem, a military problem, or a problem requiring covert action. For example, the government treated the bombing of the USS *Cole* as an international criminal investigation, even though the Yemeni government was unable or unwilling to investigate. Even today there is disagreement: witness the controversy about whether al Qaeda terrorists captured in Afghanistan should have prisoner of war status, or whether Jose Padilla (also known as Abdullah al-Mujahir), an American citizen accused of planning a radiological bomb attack, should have the legal protections of an accused criminal. There will often be disagreement on these questions, but there should be a single official responsible for deciding which course to take so there is accountability.

Conclusion

The goal is to create a "safe zone" for collecting and sharing information. This zone is one in which federal authorities can exchange information with local governments and the private sector with limited legal and personal consequences for those who might fall into its sights. Such an entity would provide officials with information that they otherwise could not have and thus provide them a better view of the domestic battlespace.

Some of the measures that would create this safe zone have already been used by intelligence and law enforcement agencies. But to be truly successful, intelligence for homeland security probably requires a new organization. First, some of the measures necessary to create the safe zone would compromise the ability of existing intelligence and law enforcement agencies to carry out their current missions. In addition, effective implementation of these measures depends greatly on a different kind of training, incentives, and culture; it would be hard to graft these onto existing organizations.

Such an organization would not need to look like a traditional intelligence agency. Indeed, there would be many advantages to abandoning some of the essential features of the current model—a single headquarters behind a fence, where staff members assume a mystique intentionally designed to set them apart from the rest of the world. Instead such an organization should be designed to assimilate and integrate itself into American society, where its sources and users of information are. By being pushed to deal directly with citizens, businesses, and local governments, staff members would also have incentives to develop the kind of culture that Americans would be more willing to accept in their midst.[6]

Notes

1. See, for example, the strategic plan published by the CIA's Directorate of Intelligence, *Analysis: Directorate of Intelligence in the 21st Century* (1996), and the articles published by this author at about the same time, "Information Age Intelligence," *Foreign Policy* 103 (Summer 1996): 35–50, and "Technology and Intelligence Reform," *Orbis* 41 (Winter 1997): 107–19. Shortly after, Ruth A. David, the deputy director for Science and Technology, circulated "The Agile Intelligence Enterprise: Enhancing Speed, Flexibility, and Capacity through Collaborative Operations," draft memo (Directorate of Science and Technology, CIA, Summer 1997), cited in *Director of Central Intelligence Annual Report for the United States Intelligence Community* (CIA, May 1999). Also see Bruce D. Berkowitz and Allan E. Goodman, *Best Truth: Intelligence in the Information Age* (Yale University Press, 2000); and Gregory W. Treverton, *Reshaping National Intelligence for an Age of Information* (Cambridge University Press, 2001).

2. See Bruce Berkowitz, "The DI and 'IT': Failing to Keep Up with the Information Revolution," *Studies in Intelligence* 41, no. 1 (2003).

3. See 5 U.S.C. 552a.

4. P.L. 107-40.

5. For an overview, see Charles Doyle, "The USA Patriot Act: A Legal Analysis," Report RL31377 (Congressional Research Service, U.S. Library of Congress, April 15, 2002).

6. The Markle Foundation Task Force for Security in the Information Age has published two reports that illustrate in greater detail how such an organization might operate. See Baird and Barksdale, chapter 4 in this volume.

III

Technology, Security, and Liberty:
The Legal Framework

7

Communications Assistance for Law Enforcement Act: Facing the Challenge of New Technologies

LARRY THOMPSON

The passage of the USA PATRIOT Act in 2001 prompted many observers to raise concerns about the legislation's impact on civil liberties and privacy.[1] However, an equally vital issue is often overlooked: given the difficulty of keeping up with developments in communications technology, is the PATRIOT Act more bark than bite? Does the government have the ability to carry out effective and timely electronic surveillance, or in pursuing tech-savvy terrorists, will it always be one step behind?

The Promise of the PATRIOT Act

New communications technologies are rapidly being developed, challenging the ability of law enforcement and intelligence authorities to keep pace. Today, the FBI is concerned about its ability to conduct surveillance on numerous modes of communication, including

—packet-mode (digitized) communications sent over the Internet;

—Internet-based telecommunications services that use packet-mode technologies, such as Voice over Internet Protocol (VoIP);

—nontraditional wireless services, such as personal digital assistants;

—Internet hotspots with multiple access points, such as cyber cafes, that help preserve users' anonymity;

—walkie-talkie networks, such as push-to-talk devices; and

—third-party calls using calling cards and toll-free numbers.[2]

The proliferation of new technologies is not the only challenge government agencies face. The task of keeping track of terrorists is made even more daunting by the sheer volume of electronic communications. As of July 2001, an estimated 165.2 million people had home Internet access in the United States.[3] In 2002 the 2.8 billion cellular and 1.2 billion fixed-line subscribers worldwide spent some 180 billion minutes on international calls.[4]

The PATRIOT Act was intended to address these challenges by ensuring that law enforcement would have the ability to eavesdrop on the broad range of technologies that would-be lawbreakers can use. As President George W. Bush announced in signing the bill, "The new law . . . will allow surveillance of all communications used by terrorists, including e-mails, the Internet, and cell phones."[5]

The electronic surveillance provisions of the new legislation offered great promise. As an important law enforcement technique, electronic surveillance allows the authorities to locate terrorists and keep track of their plans through various means, including wiretaps. The PATRIOT Act was important to electronic surveillance because Congress believed it would allow law enforcement to catch up with the new technologies used by terrorists and other criminals. As Attorney General John Ashcroft testified before Congress, "Law enforcement tools created decades ago were crafted for rotary telephone—not e-mail, the Internet, mobile communications, and voice mail."[6]

The PATRIOT Act contains several provisions that enable enhanced law enforcement authority to conduct electronic surveillance of suspected terrorists. For example, section 216 authorizes courts to grant pen register and trap-and-trace orders that are valid anywhere in the United States, rather than limiting them to the authorizing court's jurisdiction, as was previously the case. It also allows such orders, which make it possible to capture "dialing, routing, addressing, or signaling" information, to be granted for technologies other than traditional telephone lines, such as the Internet.[7] These provisions should significantly strengthen law enforcement's hand in the war on terror.

However, in crafting the PATRIOT Act, Congress was also sensitive to a modern-day dilemma: on the one hand, law enforcement needs up-to-date authority to keep pace with technological advances; on the other hand, government needs to be subject to certain checks that address important civil liberty and privacy concerns. Reflecting such concerns, section 216 does not allow authorities to intercept the content of any communications.[8] Similarly, the PATRIOT Act requires the government to make special, ex parte reports regarding the use of pen register or trap-and-trace devices to the court that issued the order within a period of thirty days.[9] As part of such reports, the relevant agency must provide the identity of the officer or officers who installed the device, the date and time the device was installed, the configuration of the device, and any information collected by the device.[10] Commentators have noted that this provision was designed to monitor the government's use of "Carnivore"-like devices and help balance security and privacy concerns.[11]

Communications Assistance for Law Enforcement Act: Hopes Unfulfilled

However, these concerns may turn out to be moot. The government may simply not have the technological ability or capacity to undertake timely electronic surveillance. This frightening possibility was contemplated by Congress when it enacted the Communications Assistance for Law Enforcement Act (CALEA).[12] Adopted in 1994, CALEA became the law because of concerns that advances in telecommunications strategy could limit the effectiveness of lawful electronic surveillance.

CALEA did not give law enforcement any new or augmented authority to conduct court-ordered electronic surveillance. This authority was already provided in the context of criminal investigations by Title III of the Omnibus Crime Control and Safe Streets Act of 1969 and by the Electronic Communications Privacy Act of 1986, which extended authorized lawful electronic surveillance to communications transmitted by wireless technology, including e-mail, data transmissions, faxes, cellular telephone, and paging devices.[13] In addition, the Foreign Intelligence Surveillance Act of 1978 authorized the issuance of electronic surveillance orders against "foreign powers" and "agents of a foreign power," including "any person . . . who . . . knowingly engages in sabotage or international terrorism, of activities that are in preparation therefore, for or on behalf of a foreign power."[14]

CALEA's goal was simply to ensure that law enforcement would have the technical capability to conduct court-ordered electronic surveillance by requiring industry to develop intercept capabilities. Under CALEA, for example, telecommunications equipment manufacturers are required to design their equipment in a way that makes wiretapping possible.[15]

Unfortunately, CALEA has not achieved its laudable objectives. In a recent report, the Department of Justice's inspector general found that nine years after the legislation was enacted, the ability to carry out electronic surveillance was lacking in many cases.[16] The inspector general's report detailed several reasons for the delays in CALEA implementation, including delays in establishing industry electronic surveillance standards through the Federal Communications Commission (FCC).[17] The report also noted that the emergence of new technologies for which electronic surveillance standards are inadequate or not yet developed will further complicate the full implementation of CALEA.[18] In concluding, the inspector general made three recommendations for improving CALEA implementation. The most important of these was for the Department of Justice to submit to Congress proposed legislation "necessary to ensure that lawful electronic surveillance is achieved expeditiously in the face of rapid change."[19]

To spur full CALEA implementation, the Department of Justice has filed a petition with the FCC asking for expedited rulemaking.[20] Among the several issues that the petition asks the commission to resolve, the most important is the Department of Justice's request that the commission find that broadband access services and broadband telephony services are subject to CALEA.

The department's petition has prompted spirited opposition, focusing largely on efforts to extend CALEA to such information services as e-mail and VoIP. As the Electronic Privacy Information Center (EPIC) noted in its comments to the FCC, the language of CALEA "unambiguously excludes information services." Consequently, EPIC argues, applying CALEA requirements to services like VoIP would "ignore the legislative intent behind the statute, and upset the delicate balance between privacy protection and law enforcement access."[21]

However well meaning, this concern is misplaced. Prior court decisions suggest that there is a strong legal case for applying CALEA to new technologies, as long as the resulting intercepts do not reveal the content of communications. In *Smith* v. *Maryland*, the Supreme Court ruled that the installation and use of a pen register was not a "search" within the meaning of the Fourth Amendment.[22] The court noted that telephone users have no justifiable expectation of privacy regarding the numbers they dial because they know that in dialing, they convey numbers to the telephone company and that the company has switching equipment for recording this information and does in fact record it for various legitimate business purposes, including printing telephone numbers on a user's telephone bill.[23] Therefore, when a telephone user dials a number, he or she voluntarily "exposes" the number to the telephone company's equipment and assumes the risk that the company might reveal the information.

Notwithstanding this Supreme Court precedent, some privacy organizations have expressed concern that pen registers and trap-and-trace orders as applied to e-mail communications might involve the interception of more than e-mail addresses. In addition, they argue that intercepted web addresses can include content information.[24] But such concerns may be moot given the circuit court ruling in *United States Telecom Association* v. *Federal Communications Commission*.[25] In denying a petition for review of an FCC order to collect information from packet-mode data pursuant to a trap-and-trace order, the circuit court noted that nothing in the order requires carriers to turn *content* information from packet-mode data over to law enforcement, absent court authorization.[26] The circuit court further pointed out that "CALEA authorizes neither the Commission nor the telecommunications industry to modify either the evidentiary standards or procedural safeguards for securing legal authorization to obtain packets from which call content has not been stripped, nor may the Commission require carriers to provide the government with information that is 'not authorized to be intercepted.'"[27] This decision makes it absolutely clear that a lawful court order must be in place before law enforcement can have access to content information from electronic surveillance.

Equally important, key provisions of CALEA are designed to safeguard privacy. For example, the legislation requires carriers to isolate the call content and call-identifying information of a target of authorized electronic surveillance from all other communications. In pleadings during the rulemaking process before the FCC, the Justice Department has noted that this requirement has very real utility in packet-mode communications, where the packets comprising the communications of a target are interspersed with other communications. If CALEA applied to broadband access or broadband telephony, carriers would be required to identify and isolate the packets of communications relating to court-authorized interceptions without reading, recording, or otherwise becoming knowledgeable of the contents of communications not subject to the order.[28] This is a very real, statutory privacy protection.

The Need for a Solution

The ambiguity concerning the scope of CALEA makes for interesting academic and policy debates, but the uncertainty is clearly detrimental to the nation's national security interests and should be resolved as soon as possible. Consider the following scenario: There is a telecommunications provider whose equipment law enforcement needs to use quickly because of information it has received regarding communications involving individuals taking part in a terrorist plot. In other words, it is the classic ticking bomb scenario. If the provider does not have adequate interception capability, a Title III or Foreign Intelligence Surveillance Act court order cannot be implemented in a timely manner. Government engineers will have to work with the provider's engineers to find a workable electronic surveillance-interception solution before any court order can be implemented. This wasted time subjects the public safety of Americans to unnecessary risk, as the Justice Department has eloquently observed:

> Today, in the context of coordinated terrorist attacks which may result in the loss of life for hundreds or thousands of Americans, any unnecessary delay is simply inexcusable. The finer nuances . . . between circuit-switched and packet-mode telephony will be lost on the surviving family members of the victims should a terrorist attack occur in the breach between the issuance of an order and its delayed implementation because of either noncoverage or noncompliance with CALEA.[29]

Court-ordered electronic surveillance is critical in the nation's efforts to prevent terrorist attacks, to catch spies, and deal with other criminals. The nation's security and public safety are simply too important to be left hanging in legal limbo. Law enforcement already has the authority to conduct electronic surveillance of all types of instrumentalities, ranging from telephone calls via regular wire facilities to VoIP. But having this authority is not sufficient. Law enforcement also needs to be able to conduct timely electronic surveillance in the face of rapidly changing technologies. To ensure that law enforcement has this ability is a national security imperative. Otherwise, terrorists will flock to those communications systems where they believe there is safety.

As the Department of Justice argues, it would be basically nonsensical to read CALEA to exempt providers of broadband access and broadband telephony services. To do so and not apply CALEA to new technologies, as the Justice Department believes Congress intended, would cause CALEA to "apply only to legacy, circuit-switched networks," and such an interpretation would render the statute obsolete.[30]

Urgent action is required to avoid this outcome. The inspector general has recommended that legislative changes be developed that are "necessary to ensure that lawful electronic surveillance is achieved expeditiously in the face of rapid technological change."[31] The Federal Bureau of Investigation is currently preparing a legislative recommendation for review by the Department of Justice and the White House. The FBI then plans to brief appropriate members of Congress on the need for a legislative remedy for delays in CALEA implementation. The FBI states that all this can be done.[32] It should and must be done. And when Congress receives the administration's proposals it should act on them with the same sense of urgency. The public safety of our nation, and even the lives of its citizens, will depend on Congress's expeditious response.

Notes

1. See, for example, American Civil Liberties Union, "Surveillance under the USA PATRIOT Act" (www.aclu.org/safefree/general/17326res20030403.html [April 3, 2003]).

2. Office of the Inspector General (OIG), *Implementation of the Communications Assistance for Law Enforcement Act by the Federal Bureau of Investigation*, OIG Audit Report 04-19 (Department of Justice, April 2004), p. ix.

3. Amitai Etzioni, *How Patriotic Is the Patriot Act? Freedom versus Security in the Age of Terrorism* (New York: Routledge, November 2004).

4. Michael A. Wertheimer, "Crippling Innovation—and Intelligence," *Washington Post*, July 21, 2004, p. A19.

5. George W. Bush, "Remarks by the President at Signing of the Patriot Act, Anti-Terrorism Legislation," October 26, 2001 (usinfo.org/wf-archive/2001/011026/epf504.htm [November 2005]).

6. Department of Justice, Office of Attorney General, "Attorney General John Ashcroft Testimony before the House Committee on the Judiciary on September

24, 2001" (www.usdoj.gov/ag/testimony/2001/agcrisisremarks9_24.htm [July 29, 2004]).

7. Tom Gede, Montgomery N. Kosma, and Arun Chandra, *White Paper on Anti-Terrorism Legislation: Surveillance and Wiretap Laws* (Washington: Federalist Society for Law and Public Policy Studies, November 2001), p. 7.

8. 18 U.S.C. 3123 and the following ones.

9. Gede, Kosma, and Chandra, *White Paper*, p. 7.

10. Ibid.

11. Ibid.

12. 47 U.S.C. 1001–21.

13. See P.L. 90-351 and 18 U.S.C. 2510–22 for the former; see P.L. 99-508 for the latter.

14. 50 U.S.C. 1801 (b)(2)(C).

15. CALEA also provides that the attorney general can reimburse telecommunications carriers for modifications to equipment, facilities, or services installed or deployed on or before January 1, 1995, to meet CALEA capability requirements. Basically, the capability requirements of CALEA require carriers to be able to isolate, intercept, and deliver communication content and call-identifying information to law enforcement pursuant to lawful government order. Carriers are responsible for modifications to equipment, facilities, and services installed or deployed after January 1, 1995.

16. OIG, *Implementation*, p. ii.

17. Ibid., p. iii. The OIG report also found that the FBI does not collect and maintain data on carrier equipment that is CALEA compliant. Therefore, although the FBI cannot ascertain the full impact of CALEA noncompliance, the bureau acknowledges that it cannot properly conduct electronic surveillance on equipment for which there is no CALEA-compliant software. The inspector general recommends that the FBI collect and maintain data on carrier equipment that is and is not CALEA compliant so that Congress and the executive branch understand the full nature and impact of delays in CALEA implementation (see pages ii and x).

18. Ibid., p. iii.

19. Ibid., p. x.

20. Department of Justice, FBI, and Drug Enforcement Administration, *Joint Petition for Expedited Rulemaking, in the Matter of Joint Petition for Rulemaking to Resolve Various Outstanding Issues Concerning the Implementation of the Communications Assistance for Law Enforcement Act*, March 10, 2004 (www.askcalea.org/docs/20040310.calea.jper.pdf [July 29, 2004]).

21. Center for Democracy and Technology, *Comments of the Electronic Privacy Information Center to the Federal Communications Commission*, April 12, 2004, pp. 2–3 (www.cdt.org/digi_tele/20040412epiccaleacomments.pdf [July 29, 2004]).

22. *Smith v. Maryland*, 442 U.S. 735 (1979).

23. Ibid., p. 741.

24. Gede, Kosma, and Chandra, *White Paper*, p. 8.

25. *United States Telecom Association* v. *Federal Communications Commission*, 227 F.3d 450 (D.C. Cir. 2000).

26. Ibid., p. 465.

27. Ibid.

28. Department of Justice, FBI, and Drug Enforcement Agency, *Joint Reply Comments of the United States Department of Justice, Federal Bureau of Investigation and Drug Enforcement Administration*, RM-10865, April 27, 2004 (www.askcalea.com/docs/20040427_jper_reply.pdf [November 2005]), p. 25, footnote 62.

29. Ibid., p. 22.

30. Ibid., p. 30.

31. OIG, *Implementation*, p. x.

32. Ibid., appendix: "FBI Response to OIG Recommendations."

8

Security, Privacy, and Government Access to Commercial Data

JERRY BERMAN

In an effort to make better use of information technology in combating terrorism, the federal government is researching, and in some cases already implementing, new ways to use the vast databases of personal information that are collected by companies in almost every line of business today and by the data aggregators that have become an important part of the information-based economy. The new data environment is remarkable for the depth of personally identifiable information that is available from private sources, as well as for the analytic capabilities to draw patterns and inferences from these data. But these opportunities also raise important concerns. This chapter focuses on the privacy issues posed by uses of private sector databases for national security: What are the risks to privacy? Why are the current privacy laws insufficient? And what rules should guide the government's use of these new capabilities?[1]

Research for this article was supported by grants from the Markle Foundation and the Open Society Institute.

The Need for Checks and Balances

Even proponents of the use of private databases for counterterrorism acknowledge the grave privacy and due process implications that arise from the new combination of vast stores of personal information and the computing power to aggregate and analyze it. As the September 11 Commission cautioned in its 2004 report, additional checks and balances are needed as the government moves forward with information sharing and other advanced technologies.[2] Yet the development and implementation of data analysis techniques for counterterrorism purposes is proceeding without a coherent legal framework. The current rules for the government's use of commercial data are fragmentary, incomplete, and unresponsive to the uses that are associated with the current emphasis on preventing terrorism through intelligence collection and analysis.

Some of the issues surrounding governmental use of commercial databases were brought to the fore in late 2002 and early 2003 thanks to news reports and commentary about the Pentagon's Total Information Awareness program. Around that same time, concerns were raised about the Transportation Security Administration's Computer-Assisted Passenger Prescreening System II project, which was supposed to use commercial and governmental databases to assign airline passengers a "risk" score that would determine which passengers would be subjected to more intensive screening before boarding their flights. In the fall of 2003, Congress cut off funding for Total Information Awareness research, and in the summer of 2004, the Transportation Security Administration announced a scaled-back passenger screening program that does not include a "risk" assessment component. But various other government agencies have continued to explore the analysis of commercial sector data for counterterrorism purposes. In May 2004 the Government Accountability Office issued a report indicating that government agencies were conducting or planning nearly 200 data mining programs for a variety of purposes, from identifying terrorists and other criminals to managing human resources.[3]

Legal authorities have broadened as well. Changes to the Attorney General Guidelines in 2002 gave the FBI authority to engage in "data mining."[4] The Information Analysis and Infrastructure Protection Directorate at the Department of Homeland Security has congressional authorization to use data mining technology. The Intelligence Reform and

Terrorism Prevention Act of 2004, along with a series of executive orders issued in August 2004, mandated creation of a broad information-sharing environment to facilitate data sharing among agencies, including data from the private sector.[5]

Developing a comprehensive set of privacy rules for programs like these not only can serve to protect individual rights but also can improve the reliability of decisionmaking, direct limited government resources toward the most fruitful lines of investigation, and ensure that the government is relying on the most accurate and complete information available.

Efficacy and Privacy

In addressing the privacy concerns associated with using information technologies to make decisions about people for counterterrorism or any other purpose, the first issue to consider is efficacy. Indeed, a fundamental privacy principle is that no more information should be collected than is necessary to achieve a legitimate purpose. If it cannot be shown that a particular use of data will yield particular improvements in national security, then the application should not be deployed, thus obviating other privacy questions.

Efficacy cannot be assessed in the abstract; policymakers must focus on specific kinds of data and specific uses. Clearly, there are some uses of commercial data that would aid in the fight against terrorism. For example, using commercially compiled address information to quickly determine where a suspect may be residing seems clearly effective. The value of other uses, such as some of the pattern-based searches sometimes referred to as data mining, remains speculative and unproven. Pattern analysis is worth studying, but until it can be proven effective, it should not be implemented. There is no set of privacy rules that will make an ineffective information program acceptable.

Effectiveness and privacy are not at odds. To the contrary, performing a privacy analysis at the time a program is being developed will help answer key questions about effectiveness: what information is being collected, with whom is it being shared, and for what purpose. If privacy is taken into account at the research and development phase, protections can be built into a system's design.

What Do We Mean by "Privacy?"

In the context of personally identifiable information provided to the government or generated in the course of commercial transactions, privacy is not just about keeping information confidential. Rather, the concept of privacy extends to information that an individual has disclosed to another in the course of a commercial or governmental transaction and even to data that are publicly available. Data privacy is based on the premise that individuals should retain some control over the use of information about themselves and should be able to manage the consequences of others' use of that information. In this sense, privacy is about limits, reliability, and consequences rather than about what is hidden.

Under U.S. privacy law as it exists today, data that are sold or exchanged commercially or that are "publicly available" are subject to statutory rules intended to protect the data subject. Arrest records, for example, are publicly available governmental records, but under antidiscrimination laws they cannot be used for employment purposes unless they include conviction or acquittal data. Under the Drivers Privacy Protection Act, driver's license data can be obtained for some purposes and not for others.[6] Bankruptcy records are publicly available, but under the Fair Credit Reporting Act (FCRA), they cannot be included in credit reports if they are more than ten years old.[7] The FCRA imposes a number of data quality requirements on commercial compilations of publicly available data that are used for certain purposes that have significant consequences for individuals, such as decisions about credit, employment, and insurance. Under the act, for example, individuals are legally entitled to access their credit reports and insist upon corrections, even though none of the data in the reports are confidential and some of them are publicly available.

The Supreme Court has recognized that the compilation of publicly available data into computerized form heightens its sensitivity. Rejecting the "cramped notion of personal privacy" that public availability of information reduces to zero one's privacy interest in disclosure of those events, the Court has held that the government can keep confidential its compilations of data made up entirely of information that is publicly available.[8] As the Court has explained, "Plainly there is a vast difference between the public records that might be found after a diligent search of courthouse files, county archives, and local police stations throughout the country and a computerized summary located in a single clearinghouse of information."[9]

A Patchwork of Laws

Government officials defending the use of commercial databases for counterterrorism purposes have frequently argued that all such uses will be in strict compliance with applicable privacy laws. Such assurances are misleading, for there are in fact very few privacy laws applicable to the government's acquisition and use of commercially compiled data for counterterrorism purposes. Ironically, under current law the private sector is actually subject to clearer and stricter statutory rules for the use of commercial data than are government counterterrorism agencies.[10]

Briefly summarized, the legal landscape is as follows: The federal Privacy Act regulates the government's use of government-controlled databases, but neither the Privacy Act nor the Constitution limits the government's access to commercial data generated in the course of business transactions.[11] In consequence, a great deal of information is available to law enforcement and intelligence agencies pursuant to voluntary disclosure or for purchase from data aggregators. "Sectoral" privacy laws do exist for specific categories of records (credit, medical, and financial), but they are riddled with exceptions of varying breadth that allow access to data for law enforcement and intelligence purposes. For example, all the privacy laws include exceptions for access pursuant to grand jury subpoena, a broad-reaching device. Under the USA PATRIOT Act, these exceptions were expanded further.

Moreover, once commercial data are obtained for counterterrorism purposes, there are few constraints on its redisclosure to other agencies for counterterrorism purposes. The Privacy Act's "routine use" exception has always allowed a good deal of sharing. This was exacerbated by the PATRIOT Act's loosening of rules for information sharing among law enforcement and intelligence agencies and the broader post–September 11 push to break down the barriers between them. Section 203 of the PATRIOT Act allows the sharing of information obtained in criminal investigations with "any other federal law enforcement, intelligence, protective, immigration, national defense, or national security official." Section 905 goes one step further and *requires* that the attorney general and the head of any other law enforcement agency disclose to the CIA all "foreign intelligence" obtained in any criminal investigation.

The combination of deep commercial stores of data and broad government authority to access them cries out for a robust framework of guidelines and accountability. Key questions include:

—When and under what circumstances should the government be able to access entire databases?

—Should there be different access standards for information of varying sensitivity?

—Should there be different standards based on the type of search: one standard for subject-based queries and a different one for pattern-based searches?

—Does the searching of databases without the disclosure of identity mitigate privacy concerns?

—If it is possible to search data without disclosing identity, what should be the rules for disclosure to the government of the identity of those whose data fit a pattern?

—Are some uses of commercially available data of less concern than others—looking up an address versus identifying a pattern that will trigger the initiation of an investigation versus a screening application that will result in the denial of a right or privilege (such as getting on an airplane)?

—Who should make various approval decisions—about access, about the patterns that are the basis for scans of private databases, and about actions taken on the basis of information?

—How can data accuracy be improved and enforced?

—When the government draws conclusions based on pattern analysis, how should those conclusions be interpreted?

—How should such conclusions be disseminated and when can they be acted upon?

—What due process rights should apply when adverse action is taken against a person based on data analysis?

Sources of Guidance

Two sets of principles can help answer these questions. The first consists of constitutional limits on the government's use of information. The second is the long-accepted set of principles known as fair information practices, which have been embodied in varying degrees in past privacy legislation.

Constitutional Guidelines

The Constitution (as currently interpreted) places few limits on government access to and sharing of commercial data. However, the Bill of Rights does impose some due process constraints on the government's use of information to make decisions affecting individuals. For example, there are strict due process protections on the use of information in criminal trials.[12]

Even before the adversarial process begins, the Constitution constrains government dependence on data of dubious accuracy. The leading case is *Arizona v. Evans*, in which a police officer ran an individual's name through his patrol car computer during a routine traffic stop.[13] The computer indicated—incorrectly—that there was an outstanding warrant for the individual's arrest. In fact, the warrant had been quashed weeks before, but the system had not been updated to reflect that. The police officer placed the individual under arrest. The Supreme Court found that the officer's action was lawful, but Justice Sandra Day O'Connor, in a concurrence joined by Justices David Souter and Stephen Breyer, said that arrests can constitutionally be made on the basis of computer matches only if it was reasonable to rely on the information in the database, and the question of reasonableness turns on whether information in the database is known to be updated and accurate.

Evans provides a constitutional basis for the principle that the government should not rely on commercial databases to arrest or detain individuals unless those databases and the method of searching them are accurate. For example, if a detention at an airport for more intensive scrutiny under a passenger screening program is considered a seizure for Fourth Amendment purposes, then the use of inaccurate or unreliable data to make that detention would not be reasonable. In the context of pattern-based searches, it seems especially true that verifying and confirming the relevance of the pattern and the reliability of the "hit" before action is taken will determine whether it is reasonable for the government to rely on pattern analysis to take adverse action against an individual.

The Constitution also limits the government's discretion to act on the basis of inaccurate or incomplete information in other areas. The government is constrained by the due process clause from denying a license to someone based on a mere arrest—that is, on incomplete information. And government decisions affecting individuals in the context of the welfare system are subject to due process protections. The pronouncement of the Supreme Court in one of the leading procedural due process cases seems relevant to the uses of information for security screening purposes: "Certain principles have remained relatively immutable in our jurisprudence. One of these is that where governmental action seriously injures an individual, and the reasonableness of the action depends on fact findings, the evidence used to prove the government's case must be disclosed to the individual so that he has an opportunity to show that it is untrue."[14]

Yet another area that is relevant to this discussion is the body of case law and administrative practice that was developed during the cold war over the extent to which due process principles are applicable to government security decisions affecting employment in the private sector. The constitutional law surrounding these and other kinds of adverse actions taken by the government might be further plumbed to discern the outer limits on the government's discretion in taking adverse action against individuals in the name of preventing terrorism.

Fair Information Practices

The principles known as fair information practices have been embodied to varying degrees in the Privacy Act, the Fair Credit Reporting Act, and the other "sectoral" federal privacy laws that govern commercial uses of information. As explained above, these laws have been riddled with exceptions (including major exceptions adopted without serious consideration in the haste to pass the PATRIOT Act) so that they do not apply in their precise terms to the new counterterrorism uses of information. Those exceptions should be revisited, for the fair information principles are remarkably relevant despite the dramatic changes in information technology that have occurred since they were first developed. While there are challenges in adapting these principles to the models of data analysis currently being pursued for counterterrorism purposes, and while some of them are clearly inapplicable to the needs of law enforcement and intelligence agencies, they provide a sound basis for thinking through the issues associated with creating a system of checks and balances for the use of commercial databases in counterterrorism activities.

NOTICE. A fundamental principle of fair information practices is that individuals should have notice both of the fact that information is being collected about them and of the purpose for which it is being collected. Notice before data collection affords individuals the opportunity to choose not to disclose the information (albeit often at the cost of forgoing the opportunity to engage in the transaction that is made conditional upon disclosure of the information). Notice also is a precondition to the individual's ability to ensure that the data being collected are accurate.

While notice may not be practical when law enforcement and intelligence agencies are collecting information from third parties in the context of an investigation focused on specific individuals, the same concerns may not apply to the newer screening uses proposed for commercial and governmental databases. In the screening context, where all individuals are subject to the same scrutiny, the government can give notice of the types of information that it is collecting or accessing and the ways in which it uses them without compromising specific investigations. The notice principle can also be implemented by informing Congress of data analysis practices.

COLLECTION (ACCESS) LIMITATION. The collection limitation principle holds that no more data should be collected or accessed than is necessary to accomplish the legitimate purpose at hand. That is, one must be able to justify the data collection by its demonstrable relevance to a defined mission. One way to implement this minimization principle in the counterterrorism context is to leave commercial information in the hands of data aggregators and have them only respond to specific queries. Another way to limit the amount of personally identifiable information collected by the government is to anonymize the data so they can be analyzed without disclosing identity. The process could have two steps, such that after the analysis had been conducted and it was determined that there was information matching the search criteria, further authority (judicial or otherwise) could be sought to obtain the identities of only those individuals who appear to have been behaving in a suspicious way. Under this concept of selective revelation, the vast majority of data would never be accessible in a personally identifiable format.

USE AND DISCLOSURE LIMITATION. As a general rule, information collected for one purpose should not be used for another purpose without consent. The use of commercial data for counterterrorism purposes necessarily involves using information for a purpose other than the one for which it was initially collected. However, the use and disclosure limitation can and should be applied to tertiary uses. Any government agency using commercial data for counterterrorism purposes should ensure that its uses of the information and its redisclosure of that information to other agencies are limited to those counterterrorism purposes. The principle of limited reuse and redisclosure is a limit on "mission creep." The principle has the added advantage of promoting data quality, by requiring the second government agency seeking the same information for an unrelated purpose to return to the source of the information, where it may have been updated or corrected.

RETENTION LIMITATION. As a general rule, data should be retained for no longer than is necessary for the purpose for which they were collected. This concept is inapplicable to some counterterrorism purposes, since the value of some investigative information may become apparent only years after it was collected. However, the retention limitation principle is relevant to various screening applications, where information is being collected on large numbers of persons without individual suspicion, as well as to other counterterrorism uses of information. Purging data significantly reduces the opportunity for abuse. Each agency that uses commercial data should have a policy delineating precisely what data it retains and for how long—and should not retain any data not necessary to the purpose for which they were collected.

DATA QUALITY. Data quality will be an essential factor in any counterterrorism use of personally identifiable information, seriously affecting both the efficacy of the program and its impact on privacy values. Before entering into a contract for access to a particular commercial database, the government should formally assess the accuracy, completeness, and timeliness of the data. Agencies should set quality standards for data acquired from commercial providers and make data quality a competitive factor in choosing among various commercial databases. Those who use the database should be apprised as to how reliable it is in order to understand how much faith to place in it. In fact, an assessment of reliability could accompany the various intelligence outcomes and products based on commercial data (just as other assessments include some indication of reliability). Agencies should insist upon periodic auditing of data quality, with negative consequences up to and including contract termination when data fail to meet agreed-upon standards. If agencies are acquiring information and bringing it into their own databases at a particular point in time, they should track the date they acquired the information and make provisions for updating it.

INDIVIDUAL ACCESS. A crucial principle found in both the Privacy Act and the laws affecting the commercial world is that individuals should have access to personally identifiable information held about them. This right is often the key to enforcing other principles. For example, the Fair Credit Reporting Act gives consumers access to their credit reports and requires credit reporting companies to correct errors. The right to insist that information about oneself be accurate is meaningless without the right to access and review the data.

In the intelligence and law enforcement contexts, there are obvious security concerns that preclude allowing terrorist suspects to review what the government knows about them. But no such risk would be posed by giving all individuals access to the commercial data about themselves that the government uses for counterterrorism purposes. Laws should be established or strengthened allowing individuals to review and verify commercial data used by the government for counterterrorism purposes. One way to do this is with legislation extending FCRA protections to commercial data used by the government for counterterrorism purposes. Even in the absence of legislative action, government agencies entering into contracts for commercial data could require as a condition of the contracts that the data aggregators provide FCRA-like rights to individuals.

In cases where individual review is not feasible, the review rights could be vested in an agency privacy officer, an inspector general, or a judge—someone who would have the authority on behalf of an individual to review records held about that person to determine if they are properly acquired, accurate, and maintained pursuant to proper authority. While such a reviewer might be able to disclose little to the individual, he could provide an oversight mechanism.

ACCOUNTABILITY AND REDRESS. Any system for acquiring personally identifiable information must have accountability and redress mechanisms. There is broad agreement that these should include audit trails to protect against unauthorized access, disclosure, or misuse. Accountability mechanisms should also include periodic inspections to ensure that privacy principles are being enforced, with follow-up reports to Congress. Any agency that uses commercial data for law enforcement or intelligence purposes should have a high-level privacy officer responsible for establishing, reviewing, and enforcing privacy standards, similar to the position created by statute within the Homeland Security Department, and an official with authority to investigate specific complaints of abuse or error. Agencies also should establish complaint procedures that include access to information for individuals who face adverse consequences (such as not being permitted to board an airplane) due to the use of commercial data.

A New Framework for Government Data Access and Use

By building on these principles, policymakers can design a more calibrated approach to government use of commercial data, one that provides guidelines for counterterrorism access to and use of commercial data, depending on the sensitivity of the information, whether it is widely publicly available, and how the government plans to use it. Such guidelines must spell out under what justification the government can obtain and use data. And searches without particularized suspicion pose special risks, requiring special rules.

Limiting Access

Some kinds of nongovernmental data should be available to analysts and investigators on a routine basis, so that they can be searched instantly, without prior approval of each query, whereas more sensitive data or data not widely available should be harder for the government to access. Name, address, and listed telephone number are examples of information in commercial databases that should be available to every law enforcement and intelligence agency with no prior authorization (although only as part of an ongoing investigation, subject to audits and other internal controls). Further study and debate would likely suggest other categories of public record information available to the public without fee (for example, licenses, property ownership records, court documents) that should be available to government investigators from commercial data aggregators without prior approval. By contrast, nonpublic commercially compiled data such as medical or financial records, travel histories, and store purchasing records should be obtained only with a court order based on an appropriate factual showing of need.

The government's authority to access information also should vary depending on how it plans to use a particular database. First, subject-based searches—where the government is seeking information on a specific individual based on particularized suspicion—should be preferred. If an FBI agent seeks access to a data aggregator's information to obtain additional data on a single suspected terrorist, that subject-based search has quite different privacy implications than a screening application or a broad pattern-based search of that same database, because screening functions and pattern-based searches review information about many people under no suspicion at all and create a far greater likelihood of false positives. Searches based on individualized suspicion are consistent with traditional investigatory practices and therefore are already subject to longstanding rules. (The weakening of some of those rules in the PATRIOT Act needs to be reconsidered.)

Second, screening programs review information about *all* individuals seeking to exercise certain rights or privileges—from gaining employment at nuclear facilities to flying on airplanes—to determine if they pose a threat. Screening is by its nature not based on individualized suspicion. Accordingly, the government should not undertake screening programs using commercial data unless specifically authorized by Congress in response to a documented risk. Information should not be retained after the screening is performed. Redress mechanisms should be established for people who believe they have been wrongly denied a right or privilege.

Third, for pattern-based searches, where information about hundreds of millions of people is searched for patterns indicative of wrongdoing, new safeguards and privacy protections are needed. Our current legal structure does not provide guidance for this new concept. Before an agent undertakes any pattern-based search, special judicial or senior executive approval should be required to ensure that there is a sufficient factual basis to design an effective pattern-based search. Any pattern-based searches conducted in the absence of specific intelligence should be explicitly approved by Congress after a demonstration of effectiveness and should be subject to accountability and redress mechanisms.

One approach for authorizing pattern-based searches of commercial data might be found in what could be called "section 215 with teeth." Section 215 of the PATRIOT Act authorizes the FBI to obtain a court order compelling disclosure of information from commercial entities. As adopted, the provision authorizes orders based solely on an assertion that the information is "sought for" an authorized counterintelligence investigation, without any factual showing. The provision seems to allow the government to compel disclosure of entire databases. It is one of the most criticized provisions of the PATRIOT Act. However, section 215 could be amended to provide a more appropriate basis for pattern-based queries. An amended section 215 could require for disclosure a finding, based on facts shown by the government, that there is reason to believe that terrorist activity fitting a certain pattern is afoot and that reliable information relevant to the interdiction of that activity would likely be obtained from the search of one or more commercial databases. The decisionmaking authority could reside in a judge or even in some independent privacy officer or board inside the relevant executive agency. Under this approach an agency that had intelligence information about a possible future attack and wanted to run a pattern-based search to identify potential planners would be required to provide the court or other independent decisionmaker

—facts reasonably indicating that a threat existed displaying certain characteristics;

—a description of the databases that the government wanted to search, including an assessment of the sensitivity of the data involved and their accuracy and reliability;

—an explanation why other methods of investigation were inadequate; and

—a statement indicating whether the commercial databases would remain under the control of the commercial source or whether they would be acquired by the government.

The court or executive authority would evaluate the application to determine if the government had shown that there were specific and articulable facts substantiating that a pattern-based search would turn up information relevant to a counterterrorism investigation, and that the databases to be searched were of such reliability and accuracy as to be likely to produce relevant information that other investigative methods would not. The process is not that dissimilar from the type of assessment a judge makes on an ordinary search warrant application.

Such a scheme might also require a government agent to obtain a judicial order after the search has been conducted, in order to find out the identities of those individuals who met the search criteria. As technology advances it may become possible to run pattern-based searches on anonymized data—that is, data that have been unconnected from the identity of the individual to whom they pertain. An agent could run a search and find out that there are ten people who fit a particular intelligence-based pattern and warrant further investigation. To find out the identity of the individual "hits," the agent would have to return to court to justify the need for the information based on the likelihood that the pattern demonstrates that the individuals are terrorists.

This concept of selective revelation—anonymized searches combined with judicial approval for access to identifying information—was embraced by the Defense Department's Technology and Privacy Advisory Committee, created in 2003 to address concerns about the Total Information Awareness program. The committee issued its recommendations for data mining rules in March 2004.[15] It also proposed that data mining searches using personally identifiable information that is likely to concern U.S. persons should be subject to the prior approval of the Foreign Intelligence Surveillance Court, which has the authority to issue wiretap and other surveillance orders for intelligence investigations occurring within the United States.

Where searches are conducted using large amounts of data concerning mostly innocent people, it is preferable to leave the data with the entity that collected them, rather than bringing the data into the government's files. If the data are transferred to the government, nothing should be retained after the pattern-based search other than data indicating specific suspicion.

Providing Mechanisms for Redress

Traditionally, government counterterrorism uses of personally identifiable information were fairly narrow. In the case of citizens and permanent resident aliens, the government's only active option was criminal prosecution. In the criminal justice system, the rules for use of information were clear and protective of the individual. In the new environment, with its emphasis on prevention outside the criminal justice system, the government is likely to take a range of adverse actions against citizens for which the due process rules are unclear.

A full discussion of consequences and due process is hampered by the lack of specificity in many discussions about how the government will be using commercial information and other data analysis techniques. How will "knowledge" generated by computerized analysis of commercial and government data be used? Could it trigger a criminal or intelligence investigation? Will it be used to build a criminal case? (Once a criminal investigation proceeds to the stage of search and seizure or arrest, traditional probable cause protections come into play.) Will it be used to place someone on a watch list? Will it be used for screening purposes—to trigger a more intensive search of someone seeking to board an airplane, to keep a person off an airplane, to deny a person access to a government building, to deny a person a job? It is unclear exactly what are the means of preventing terrorism outside the criminal justice and immigration systems, but it is clear that personally identifiable information will be used to screen individuals at airports and possibly in other transportation and employment contexts.

Given the new emphasis on screening and prevention, the criminal due process rules will often be inapplicable. The establishment of access and approval standards for pattern-based searches, the use of anonymization techniques to shield the identity of persons in databases, and auditing to identify unauthorized uses only partly resolve the concerns raised by the use of commercial data. Even if all those techniques are used, the question remains: how will "hits" be used, and what opportunity will an individual have to prove that he or she is not a risk?

Providing due process to those who face adverse consequences involves protecting people not only against the consequences of abuse but also against authorized uses that happen to be mistaken, either because they were based on erroneous information or because the analysis results in false positives. Different rules will probably be needed depending on the different consequences; a person denied a job may have more rights than a person subject to a baggage search at an airport. For example, one possible rule could be that further traditional investigation is required before any overt action can be taken against an individual based solely on a pattern-based query. If the process is of a screening nature, such as the screening of airline passengers to determine whether any are on a terrorist watch list, the set of procedures laid out in the Privacy Act or the Fair Credit Reporting Act (with some modifications given the national security context) may serve as a model.

For other uses it will be necessary to create new control and oversight mechanisms. For example, if a pattern-based search leads to criminal or intelligence investigations, prompt resolution of those investigations would be desirable, so that those upon whom the computer mistakenly cast suspicion do not remain under surveillance. Current guidelines for FBI investigations set various time limits beyond which investigative activity cannot proceed without headquarters or Department of Justice approval. For investigations opened as a result of pattern-based searches, those timelines should be shorter than normal, to promptly resolve suspicions.

Conclusion

Under existing law the government can ask for, purchase, or demand access to most private sector data subject to few limits. Sharing is generally broadly permitted among agencies with counterterrorism responsibilities. Constraints on how the government can use the data once accessed are fragmentary. This has produced a situation of uncertainty, not only for a public wary of government overreaching but also within both the government and the private sector. The current lack of clarity is holding back efforts to develop potentially useful applications of commercial information. There is an urgent need for structures to fill this gap.

Given the current emphasis on prevention and screening outside the criminal justice context, governmental use of commercial data presents new challenges that the current legal structure does not adequately address. A new framework is needed. However, it is not necessary to design that framework in isolation. It can draw upon existing constitutional doctrine and the principles of fair information practices. And best of all, rules that guide government use of information by making decisions more reliable, transparent, and accountable will not only protect civil liberties but will also enhance the effectiveness of government counterterrorism activities.

Notes

1. The ideas in this article have been informed by the deliberations of the Markle Task Force on National Security in the Information Age and are based in many respects on the Task Force's reports. See Markle Foundation Task Force on National Security in the Information Age, *Protecting America's Freedom in the Information Age* (www.markle.org/downloadable_assets/nstf_full.pdf [October 2002]), and *Creating a Trusted Network for Homeland Security* (www.markle.org/downloadable_assets/nstf_report2_full_report.pdf [December 2003]). However, in this article the author does not speak for the Markle Task Force.

2. *The 9/11 Commission Report: Final Report of the National Commission on Terrorist Attacks upon the United States* (Government Printing Office, July 22, 2004).

3. General Accounting Office, *Data Mining: Federal Efforts Cover a Wide Range of Uses*, GAO-04-548 (2004).

4. John Ashcroft, "The Attorney General's Guidelines on General Crimes, Racketeering Enterprise and Terrorism Enterprise Investigations," May 30, 2002 (www.usdoj.gov/olp/generalcrimes2.pdf [November 2005]).

5. See P.L. 108-458, and Executive Order 13353, *Establishing the President's Board on Safeguarding Americans' Civil Liberties*; Executive Order 13354, *National Counterterrorism Center*; Executive Order 13355, *Strengthened Management of the Intelligence Community*; and Executive Order 13356, *Strengthening the Sharing of Terrorism Information to Protect American*, all issued August 27, 2004 (www.whitehouse.gov/news/orders/ [November 2005]).

6. P.L. 103-322.

7. 15 U.S.C. 1681.

8. *Department of Justice* v. *Reporters Committee*, 489 U.S. 749, 762–63 (1989).

9. Ibid., p. 764.

10. The landscape of legal rules is laid out in two charts prepared for the Markle Foundation Task Force by the Center for Democracy and Technology, "Commercial Access to Information" and "Law Enforcement and Intelligence Access to Information" (www.cdt.org/security/guidelines/ [November 2005]).

11. 5 U.S.C. 552a.

12. A central principle of the criminal justice process is the right to confront evidence, challenge its accuracy, and offer countervailing information. Some criminal due process rules prohibit the government from even introducing incomplete evidence. For example, one form of publicly available information offered by commercial databases—the arrest record unaccompanied by information of the disposition of the charge—is generally inadmissible in criminal trials.

13. *Arizona* v. *Evans*, 514 U.S. 1 (1995).

14. *Goldberg* v. *Kelly*, 397 U.S. 254, 270 (1970).

15. Technology and Privacy Advisory Committee, *Safeguarding Privacy in the Fight against Terrorism* (Department of Defense, March 2004).

9

Foreign Intelligence Surveillance Act: Has the Solution Become a Problem?

BERYL A. HOWELL

The Foreign Intelligence Surveillance Act of 1978 (FISA) was originally enacted to protect civil liberties by imposing checks and balances on the president's power to collect foreign intelligence through electronic surveillance. Recent developments have prompted concerns that FISA no longer serves this function. First, the USA PATRIOT Act's amendments to FISA have broadened the secret surveillance power of the president and the attorney general.[1] Second, the Bush administration's penchant for secrecy and resistance to congressional oversight have exacerbated concerns that FISA has routinized procedures for abusive investigations, rather than stopping them and ensuring government accountability, as originally intended.[2]

Although Congress moderated the administration's more extreme proposals to amend FISA as part of the PATRIOT Act, this law relaxed the standards and procedures for collecting foreign intelligence through electronic surveillance and physical searches, as requested by the executive

Many thanks to Simon Steele, John Elliff, and Julie Katzman for their thoughtful review and comments on this essay; any mistakes in perception or the law remain my own.

branch.[3] These statutory changes were made to facilitate information sharing among intelligence and law enforcement agencies and to relieve administrative burdens so resources could better flow to intelligence gathering and analysis.[4] Notably, the PATRIOT Act amended FISA's original requirement that collecting foreign intelligence be "the purpose" for electronic surveillance by specifying that it need only be a "significant purpose."[5] This will most likely result in more frequent use of FISA-derived evidence in criminal investigations and prosecutions.

This chapter reviews the legislative history leading to the change in FISA's "purpose" restriction and some related amendments made by the PATRIOT Act. It also explores ways to improve public confidence in the exercise of FISA authority by ensuring fairness in criminal proceedings and enhancing public accountability.

FISA's Rationale

FISA's enactment was the culmination of an eight-year debate over the scope of the president's power to engage in warrantless electronic surveillance—so-called black bag jobs—to obtain foreign intelligence information.[6] The practice of warrantless surveillance to protect national security dated back to at least the 1930s but lacked clear standards and accountability. Before FISA, the executive branch was the sole and final arbiter of the circumstances where electronic surveillance was warranted to collect foreign intelligence, subject only to sporadic review by a select group in Congress. As a result, this power was frequently abused.

The Church Committee, a congressional body set up in 1975 to investigate these issues, found that the executive branch had used national security criteria to justify the warrantless surveillance of a broad range of individuals, including "a United States Congressman, congressional staff member, journalists and newsmen, and numerous individuals and groups who engaged in no criminal activity and who posed no genuine threat to the national security, such as two White House domestic affairs advisers and an anti-American Vietnam War protest group."[7] In particular, warrantless surveillance of government employees and journalists had been conducted on a number of occasions to identify the sources of leaked classified information. This practice, along with the targeting of persons or groups who disagreed with government policies and engaged in lawful domestic political activity, raised serious civil liberties concerns.[8]

These concerns were compounded by practical questions about whether unilateral executive branch surveillance would survive court challenges without legislative authority. Judicial decisions in the 1970s curtailed the president's power to conduct certain warrantless electronic surveillance but left open the question of when exercising such power would be constitutional. For example, in *United States* v. *United States District Court* (the Keith case), the Supreme Court held unconstitutional the Nixon administration's use of electronic surveillance without a court order "to protect the nation from attempts of *domestic organizations* to attack and subvert the existing structure of the government."[9] This case left unresolved the question of whether the president has an inherent power under the Constitution's Article II to conduct warrantless surveillance to gather foreign intelligence.[10] It also left doubt about the admissibility in criminal cases of evidence derived from warrantless national security surveillances, putting at risk prosecutions that might otherwise be successful. This legal uncertainty, compounded by the public outcry regarding Hoover-era abuses, united a broad ideological spectrum behind FISA's passage. Too much was at stake to gamble on inaction.

Congress's solution in FISA was threefold. First, it imposed accountability standards on domestic surveillance of foreign powers and their agents by granting the executive branch strong national security powers while creating safeguards against abuse. Rather than permit the president to exercise surveillance powers in secret, FISA made public the standards for and limitations on such surveillance. Second, FISA limited the authorized targets of surveillance to foreign powers and their agents. This ensured that there would be no statutory basis for surveillance of domestic groups allegedly threatening national security, other than by the normal means applicable to domestic criminal investigations. Third, Congress made clear that it was not recognizing, ratifying, or denying the existence of any inherent presidential power to authorize warrantless surveillance in the United States. Rather, the enactment of FISA was intended to "moot the debate over the existence or nonexistence of this power, because no matter whether the president has this power, few have suggested that his power would be exclusive."[11]

Checks and Balances: The Role of Congressional Oversight

In order to limit executive branch discretion and the concomitant risk of abuse, FISA made the president subject to statutory standards and guidelines in conducting national security–related surveillance of foreign powers and their agents operating in the United States. Decisions to engage in such surveillance were made subject to judicial review by the Foreign Intelligence Surveillance Act Court and the FISA Court of Review and to improved congressional oversight through both secret and public reports to Congress.[12] In addition, the statute incorporated a broad "discovery" right for congressional committees. FISA expressly states that "nothing in this title shall be deemed to limit the authority and responsibility of the appropriate committees of each House of Congress to obtain such information as they may need to carry out their respective functions and duties."[13]

Under FISA, oversight responsibility is shared by multiple committees with different areas of expertise. While the Intelligence Committees have oversight responsibility over all governmental intelligence activities, the Judiciary Committees, which have responsibility for espionage, civil liberties, civil and criminal judicial proceedings, and government information, have concurrent jurisdiction over FISA. Nonetheless, the Bush administration has resisted providing the Senate and House Judiciary Committees with "such information as they may need to carry out their respective functions and duties."[14] In part as a result, Congress stipulated that virtually all FISA amendments in the PATRIOT Act would "sunset" in 2005. This provision was designed to give the Department of Justice an incentive to cooperate more fully with oversight efforts.[15]

The "Log" and the "Wall"

In addition to providing for congressional oversight, FISA included safeguards to ensure accountability among senior executive branch officials for decisions to exercise FISA power. The act required that an executive branch official, designated by the president and confirmed by the Senate, certify inter alia "that the purpose of the surveillance is to obtain foreign intelligence information."[16] Law enforcement was not a permissible purpose for initiating a FISA surveillance or search. However, FISA explicitly recognized the fact that surveillance undertaken under its auspices could yield information for use in criminal investigations.[17]

Indeed, FISA permits federal officers and employees to use and disclose FISA-derived information for any "lawful purposes" and contemplates that the government might use such information "to enter into evidence or otherwise use or disclose in any trial, hearing, or other proceeding in or before any court, department, officer, agency, regulatory body, or other authority of the United States."[18] The information's dissemination to and use by state and local authorities, as well as foreign officials, was also clearly contemplated.[19] Moreover, United States persons, including American citizens and resident aliens, are subject to FISA surveillance as "agents of a foreign power" only when they engage in certain enumerated activities involving possible criminal activity, which might lead to enforcement actions.[20] Finally, the statute sets forth detailed procedures for the use of FISA surveillance information by law enforcement, including procedures for notifying "aggrieved persons" and disclosing FISA applications and orders to a defendant in criminal proceedings.[21] In short, the act recognized that FISA-derived information could and would be used in criminal matters.

Nevertheless, FISA was not intended to be a tool for prosecutors to supplant more stringent procedures for gathering evidence in criminal prosecutions. The statute's structure reflects this understanding: the threshold certification requirement of "the purpose" for surveillance to collect foreign intelligence information is totally separate from the minimization procedures for handling and permissible uses of the FISA-derived information after collection.[22] As the 1978 Senate Select Committee on Intelligence report noted, "Although there may be cases in which information acquired from foreign intelligence surveillance will be used as evidence of a crime, these cases are expected to be relatively few in number."[23] Instead the act was intended to supplement the investigative options available to the government and "cover the situation where the government cannot establish probable cause that the foreign agent's activities involve a specific criminal act."[24]

To minimize the potential for abuse, particularly in those cases where foreign intelligence gathering and law enforcement purposes might overlap, FISA narrowly defined the categories of persons—both U.S. persons and foreigners—who were authorized targets of FISA surveillance. U.S. persons, in particular, had to be suspected of engaging in criminal activity connected to espionage, sabotage, international terrorism, or clandestine intelligence gathering on behalf of a foreign power. In addition, in 1995 the Justice Department implemented procedures to assure the FISA Court that the primary purpose of a FISA electronic surveillance or search was to collect foreign intelligence information, even when evidence of criminal activity was obtained and criminal investigators were involved.

These procedures created a "wall" between criminal investigations and intelligence gathering by excluding criminal investigators and prosecutors from the "direction and control" of FISA tools. The logic was simple, if procedurally awkward. For example, whenever the two activities overlapped, an individual who was not involved with the criminal investigation would be appointed to "review all of the raw intercepts and seized materials and pass on only that information which might be relevant evidence."[25] The Justice Department also prepared reports about "consultations and discussions between the FBI, the Criminal Division, and the U.S. Attorney's offices in cases where there were overlapping intelligence and criminal investigations or interests" and prepared a log of all contacts in a particular matter between intelligence and criminal officers within the department for the FISA Court.[26] When criminal investigations of FISA targets were being conducted concurrently, with prosecutions likely, the FISA Court itself supervised the dissemination of information to criminal investigators and, in the court's words, "became the 'wall.'"[27]

Maintaining a log of contacts and reporting to the FISA Court was a cumbersome, time-consuming, and error-prone process.[28] Compliance with these procedures did not bar communications and coordination among law enforcement and intelligence agencies, but it did require extensive monitoring and reporting. More important, screening by the "wall" may have hindered effective exploitation of all relevant surveillance leads. This was the context in which the Justice Department proposed a change in FISA's "purpose" certification requirement following the September 11 terrorist attacks.

From "the Purpose" to "a Significant Purpose"

The Bush administration originally proposed to modify FISA and relax the restriction that obtaining foreign intelligence information must be "the purpose" of FISA surveillance and searches to be only "a purpose." The administration posited that this change was necessary to facilitate information sharing among law enforcement and intelligence agencies. As the Justice Department explained,

> Current law requires that FISA be used only where foreign intelligence gathering is the sole or primary purpose of the investigation.

This section will clarify that the certification of a FISA request is supportable where foreign intelligence gathering is "a" purpose of the investigation. This change would eliminate the current need continually to evaluate the relative weight of criminal and intelligence purposes, and would facilitate information sharing between law enforcement and foreign intelligence authorities which is critical to the success of antiterrorism efforts.[29]

This argument was set against the backdrop of early press reports that two of the suspected hijackers had been on a terrorist watch list maintained by the CIA, which had informed the FBI only after they had entered the United States.[30]

For several significant reasons, key members of Congress were cautious, however, about expanding the use of FISA to cover any situation where "a purpose"—no matter how remote or speculative—was to obtain foreign intelligence information. First, they were concerned it would allow far broader use of FISA in run-of-the-mill criminal cases, where law enforcement would be tempted to take advantage of easier mechanisms for obtaining court authorizations as well as more stringent restrictions on disclosure that would prevent defendants from challenging the basis for surveillance.

Second, critics of the administration's proposal believed that it addressed the wrong stage of the FISA process. The administration sought to broaden the scope of permitted information gathering, but to many observers the real national security problem was not information gathering but the sharing of collected information among agencies. Indeed, even one of the Department of Justice experts on FISA stated that the requested change to "a significant purpose" was "not so much designed to expand the kinds of information that we can obtain, but rather to ensure that when we get the information, we can coordinate properly between the intelligence side and the law enforcement side of the government."[31] He explained that the change in the purpose language reflects that in many cases,

There will be law enforcement equities that are implicated by the activity that is under surveillance. We need to be able to coordinate between our law enforcement authority elements in the government and our intelligence in the government so that we can have a coherent, cohesive response to an attack like the one we experienced on September 11 and not end up in a situation where we have a splintered, fragmented approach and the left hand and the right hand don't know what each other is doing.[32]

Based in part on this reasoning, and on the concern that broadening "the purpose" language of FISA as the Bush administration requested might put "at risk the entire FISA statute," Senator Patrick Leahy proposed an alternative fix to break down "stovepipes" of information and facilitate information sharing between intelligence and law enforcement agencies.[33] Specifically, the proposal authorized intelligence officers who were initiating and conducting FISA surveillance to consult with federal law enforcement officers in order to protect the United States against actual or potential attacks, sabotage, international terrorism, or clandestine intelligence activities by foreign powers or their agents. The administration agreed to this proposal, which became the so-called coordination provision found in section 504 of the PATRIOT Act.[34] The coordination provision was designed to foster exchanges of information and expertise and improve coordination between intelligence and law enforcement, without changing "the purpose" certification language.

Finally, critics viewed as constitutionally suspect the Bush administration's proposed change to authorize FISA electronic surveillance and physical searches when "a"—not "the" or primary—purpose was to obtain foreign intelligence information.[35] In response, the administration promised an opinion letter from the Justice Department's Office of Legal Counsel justifying the department's constitutional analysis in support of the proposed change to "a purpose."[36]

This letter never materialized. Instead, at a subsequent hearing of the Senate Judiciary Committee, Attorney General John Ashcroft was asked by Senator Diane Feinstein to consider a change in "the purpose" certification language to "substantial or significant purpose" rather than the administration's broader proposed change to "a purpose."[37] Attorney General Ashcroft responded:

> I think if I were forced to say if we are going to make a change here, I think we would move toward thinking to say that if 'a purpose' isn't satisfactory, say 'a significant purpose' reflects a considered judgment that would be the kind of balancing that I think we are all looking to find. If I were to choose one of your words, I think that would be the one I would choose.[38]

Following this exchange, the Justice Department forwarded a letter from Assistant Attorney General Daniel J. Bryant defending the constitutionality of a change to "a significant purpose."[39] This was the formulation ultimately adopted as part of the PATRIOT Act.

Ironically, despite the level of criticism that has been leveled at the amendments to FISA made in the PATRIOT Act, Congress was effective in stopping the executive branch from expanding the permissible purpose of FISA to allow its use in any criminal investigation, regardless of whether that conduct was connected to foreign intelligence–related crimes or just as susceptible to investigation using normal criminal investigatory tools. Use of FISA for law enforcement purposes remains limited to foreign intelligence crimes, such as espionage, sabotage, international terrorism, and entry into the U.S. with a fraudulent identity.[40] Collection of evidence about these crimes may be the primary purpose of FISA. However, even in such cases, the collection of foreign intelligence information must also be a "significant purpose" of the surveillance. Indeed, the FISA Court of Review has made clear that when the government's primary objective is to prosecute an agent of a foreign power for a crime unrelated to foreign intelligence, this would "disqualify an application."[41] FISA powers may not "be used as a device to investigate wholly unrelated ordinary crimes."[42] The Justice Department's argument to the contrary was rejected.[43]

Additional Information-Sharing Provisions

Concerns over apparent breakdowns in information sharing also prompted the addition of several provisions designed to encourage law enforcement officials to share relevant information with intelligence officials.[44] At the behest of the Senate Select Committee on Intelligence, section 905 of the PATRIOT Act requires the attorney general and other federal law enforcement officials to "expeditiously disclose" to the director of Central Intelligence "foreign intelligence acquired by an element of" any department or agency with law enforcement responsibilities.[45] However, certain classes of foreign intelligence information may be exempt from the mandatory disclosure requirement if "the attorney general determines that disclosure of such foreign intelligence . . . would jeopardize an ongoing law enforcement investigation or impair other significant law enforcement interests."[46]

To ensure that criminal investigators and prosecutors recognize foreign information that might be of interest to intelligence agencies, section 908 of the PATRIOT Act further requires the attorney general, in consultation with the director of Central Intelligence, to develop training programs to help federal, state, and local personnel identify foreign intelligence information discovered in the normal course of their duties and to use such information when appropriate. Combined with other sections of the act that allow broader sharing of grand jury, criminal wiretap, and other criminal justice information, these training and disclosure requirements form a powerful mandate to force information sharing among law enforcement and intelligence agencies.[47]

Still, legal mandates alone cannot eliminate the cultural, behavioral, and resource obstacles to effective information sharing and analysis of foreign intelligence data. Consequently, Congress designed a number of measures included in the PATRIOT Act to address these fundamental problems. These included not only the Leahy coordination provision in section 504 but also provisions to increase the number of translators within the FBI and to improve security on the northern border. Nevertheless, practical obstacles remain, with recent reports from the Justice Department that "the intelligence collected from counterterrorism surveillance and other means is piling up, because the bureau lacks the funds and translators it needs to make sense out of them."[48]

Emerging Problems: Discovery in Criminal Cases and Mechanisms for Redress

The increased involvement of criminal investigators and prosecutors in the FISA process will likely result in earlier identification of criminal conduct and greater readiness to use this evidence in criminal prosecutions. This trend will force courts to grapple with the fairness issues posed by FISA discovery rules. In fact, it is somewhat of a misnomer to speak of "FISA discovery" since no criminal defendant has ever received a single FISA application in discovery. Normally, defendants in criminal cases are given access to complete applications for and court orders authorizing searches, wiretaps, and other forms of surveillance.[49] Yet discovery of FISA applications and orders is treated differently from discovery of all other forms of material, including other classified material.[50] This distinction is, in many ways, an accident of history.

The discovery of classified material in criminal cases is generally governed by the Classified Information Procedures Act (CIPA), which was enacted two years after FISA.[51] Congress failed to harmonize the new standards in CIPA, however, with the vague discovery process set forth in FISA. CIPA established detailed procedures for limiting access to classified information in a criminal prosecution while allowing a defendant sufficient access to challenge the constitutionality and accuracy of the evidence directed against him or her. Under CIPA a court may

—issue protective orders prohibiting defendants from disclosing classified information;

—authorize the government to delete, summarize, or substitute specified items of classified information from criminal discovery upon proper showings;

—review government submissions regarding such information on an ex parte basis;

—conduct closed hearings;

—provide advance notice to the government of any ruling requiring disclosure of classified information and an opportunity for an interlocutory appeal from such an order; and

—allow the government a chance to dismiss part or all of its case or enter into stipulations to avoid such disclosure.

In contrast to what is allowed under CIPA, judges are not permitted to disclose FISA applications, orders, "or other materials" when the attorney general asserts under oath "that disclosure . . . would harm the national security." Following in camera and ex parte review, the judge may disclose portions of the FISA materials only "where such disclosure is necessary to make an accurate determination of the legality of the surveillance."[52] The attorney general normally opposes disclosure, and judges have felt sufficiently competent to make these determinations without the necessity of disclosure.[53] Should a judge determine otherwise, any disclosure must take place "under appropriate security procedures and protective orders." Such procedures are not spelled out in the statute and remain largely undefined by the courts, primarily because there has never been a disclosure order in FISA's nearly three decades of existence.

This bifurcated discovery regime is deficient for several reasons. First, under the current system, there is a high degree of randomness in ascertaining which set of discovery rules will govern a particular piece of evidence. Whether CIPA or FISA is used depends not upon the content or sensitivity of the classified material but upon how the issue is raised in the course of litigation. The FISA discovery rule applies "whenever a motion is made" by an aggrieved person, so if a challenge to the evidence is raised by a defense motion to suppress at the pretrial stage, FISA rules control discovery. However, if the same material is discoverable, not by defense motion but by affirmative obligations on the government under Federal Rule of Criminal Procedure 16 or the Jencks Act, or as exculpatory evidence under *Brady* v. *Maryland,* then CIPA likely controls.[54] This dichotomy strains logical justification.

Second, the current FISA regime is overly restrictive. Under the FISA test, a judge does not disclose any part of a FISA application unless the judge determines that, in effect, he or she is incapable of arriving at an accurate suppression ruling without the defense attorney's help. As a result, disclosure never occurs. Courts have recognized that the need for disclosure would arise if an initial review of the FISA application revealed irregularities such as misrepresentations of fact, vague identification of the persons to be surveilled, or interception of a significant amount of nonforeign intelligence information that showed noncompliance with minimization orders.[55] Yet without being able to review the applications and orders, the defense is left to make speculative suppression arguments about what applications might say on these matters without a fair opportunity to dispute the contents. The result is that the adversary process is undermined, which is especially problematic when the question at issue is whether intelligence agents violated the constitutional rights of those whom the government seeks to imprison or even execute.

Third, it is unclear whether certain classes of material are governed by CIPA or FISA discovery rules. For instance, FISA discovery procedures apply only to "applications or orders or other materials relating to electronic surveillance" and to "physical search authority."[56] FISA is silent on whether FISA applications for orders compelling production of business records or tangible items are discoverable under CIPA or FISA rules. In addition, although the statutory language appears to imply that the fruits of a FISA order are governed by FISA discovery rules, the statute makes little sense when applied in this manner. Discovery is only theoretically granted under FISA when it is necessary to determine the legality of surveillance. That rationale, however, would almost never justify the discovery of "fruits" evidence from a physical search or electronic surveillance, or a motion based on some other grounds.

Minimizing or eliminating the differences in discovery procedures and standards between FISA-related material and all other classified information in criminal cases should be considered a priority, especially in light of fairly recent evidence of problems in FISA implementation. A declassified

FBI internal memorandum dated April 14, 2000, cites a number of errors that took place within the first quarter of that year, including videotaping a meeting without FISA authorization, wiretapping a cell phone after the target had stopped using it and the phone was reassigned to a new person, intercepting e-mail without authorization in the FISA order, unauthorized searches, incorrect addresses, and incorrect interpretations of a FISA order.[57] As more FISA-derived information is used to support criminal charges, the adversary process should be allowed to work by providing a meaningful opportunity to challenge such mistakes.

In the face of the FISA Court's findings of serious material misstatements and omissions in FISA applications, this situation is cause for concern. In theory, the adversarial process is effective at identifying errors, but if FISA applications, orders, and related material are essentially exempt from that process, any such errors may escape scrutiny. Moreover, the ability of defendants to challenge the constitutionality of FISA orders authorized under recent expansions of FISA authority, such as the so-called lone wolf amendment (extending FISA to cover suspected terrorists who are not connected to a foreign power), is seriously hampered if they are unable to review the application and order to determine whether new FISA authorities were the predicate for the evidence used against them.[58] In short, the FISA discovery rules raise serious fairness issues about how FISA-derived information and its fruits are used against defendants in criminal cases. Congress should consider and address these issues.

In addition, the evidence of errors in implementing FISA highlights the need to improve measures for redress under the PATRIOT Act. Section 223(c)(1) of this law provides a remedy for those persons who are aggrieved by willful FISA violations. It authorizes aggrieved persons to bring civil actions in U.S. district courts against the United States to recover monetary damages.[59] Moreover, when circumstances surrounding the violation raise "serious questions" about whether a government official acted willfully or intentionally, a proceeding must be promptly initiated to determine whether disciplinary action against the official is warranted as well as a referral to the agency's inspector general.[60] This statutory remedy may be illusory in practice since FISA's discovery and review procedures in 50 U.S.C. 1806(f) (electronic surveillance), 1825(g) (physical searches), and 1845 (pen registers or trap-and-trace devices) "shall be the exclusive means by which materials governed by those sections may be reviewed." This restriction raises the question of whether grievances can be effectively discovered, let alone litigated, when the aggrieved party cannot review the underlying application to show inaccuracies, falsehoods, or insufficient showings to support exercising FISA authority.

Conclusion

Americans are naturally suspicious of secret proceedings and rightfully fearful that unchecked surveillance powers will be abused. FISA was in large part a response to such fears and actual abuses. In enacting FISA, Congress sought to protect civil liberties by imposing standards and judicial and congressional accountability on national security investigations, while at the same time affording the government broad and generally secret investigative power. However, since the passage of the PATRIOT Act, the debate has shifted to focus on whether the expansion of FISA surveillance powers has given too much discretion to the executive branch and whether FISA procedures risk undermining the more stringent probable cause and notice requirements in criminal law.

Enhanced reporting requirements and vigorous oversight, combined with greater focus on ensuring fairness in the use of FISA-derived information in criminal proceedings, can do much to address these concerns. Such measures would also foster public confidence in this law, which was undermined by revelations about ongoing and serious problems in the FISA process that emerged only *after* Congress had passed the PATRIOT Act.[61]

FISA was recently amended, as part of the Intelligence Reform and Terrorism Prevention Act in December 2004, with additional reporting requirements to improve accountability in its use.[62] The attorney general is now required to report annually the aggregate number of persons targeted for multiple types of FISA orders and the number of times FISA information is authorized for use in criminal proceedings. In addition, he or she must provide to congressional oversight committees an annual summary of significant legal interpretations of FISA and of decisions or opinions of the FISA Court or FISA Court of Review.

More can and should be done, however. Additional consideration should be given to providing an expeditious mechanism within the Department of Justice for handling complaints from recipients of FISA orders and national security letters, which are forms of administrative subpoenas authorized to be issued under statutes other than FISA to compel production of information in national security–related investigations. The attorney general should be required to report the types and numbers of such complaints to appropriate oversight committees. Finally, the burden rests on the Senate and House Intelligence and Judiciary Committees to monitor such reports closely and ensure that public confidence in the proper use of FISA powers is warranted. As Senator Leahy said on the Senate floor in early 2003, "Sunlight is the best solvent for the sticky and ineffective machinery of government, and it is the best disinfectant to discourage the abuse of power."[63]

Notes

1. *Uniting and Strengthening America by Providing Appropriate Tools Required to Intercept and Obstruct Terrorism (USA PATRIOT) Act of 2001*, Public Law 56, 107 Cong. 1 sess. (October 26, 2001).

2. The current administration's propensity for "Nixon-style secrecy" has been widely noted. See, for example, John Dean, "Hiding Past and Present Presidencies: The Problems with Bush's Executive Order Burying Presidential Records" (writ.news.findlaw.com/dean/20011109.html [November 9, 2001]); Paul Krugman, "Lifting the Shroud," *New York Times*, March 23, 2004, p. A23; Murray Light, "Secrecy Is Growing under Bush," *Buffalo News*, December 28, 2003, p. F5; Simon English, "Cheney Fights to Keep 'Enron Papers' Secret," *Daily Telegraph*, December 16, 2003, p. 29.

3. Even after passage of the USA PATRIOT Act, the administration has obtained additional FISA changes. See, for example, *Intelligence Authorization Act for Fiscal Year 2002*, Public Law 108, 107 Cong. 1 sess. (December 28, 2001), sec. 314, "Technical Amendments"; *Intelligence Reform and Terrorism Prevention Act of 2004*, Public Law 458, 108 Cong. 2 sess. (December 17, 2004), sec. 6001, "Individual Terrorists as Agents of Foreign Powers." For a review of other amendments to FISA and the federal criminal code made in the PATRIOT Act, see Beryl A. Howell, "USA PATRIOT Act: Seven Weeks in the Making," *George Washington Law Review* 72, no. 6 (2004): 1145–1207.

4. The PATRIOT Act contained the following amendments to FISA: Section 206 provides roving electronic surveillance authority under FISA where the court finds that the actions of the target may have the effect of thwarting the identification of a specified person. Section 207 extends the duration of FISA surveillance of non–United States persons who are agents of a foreign power from 90 to 120 days, and the period for extensions from 90 days to one year. Section 208 increases the number of federal district judges designated to serve on the FISA Court from seven to eleven and requires that no fewer than three of the judges reside within twenty miles of the District of Columbia. Section 214 expands pen register and trap-and-trace authority under FISA by eliminating the required

showing that the target is in contact with an "agent of a foreign power." Section 215 removes the "agent of a foreign power" standard for court-ordered access to certain business records under FISA and expands the types of records and tangible items subject to court-ordered access. Section 218 requires certification that "a significant purpose" rather than "the purpose" of a surveillance or search under FISA is to obtain foreign intelligence information. Section 223 creates civil liability for violations, including unauthorized disclosures by law enforcement authorities of FISA information. Section 224 applies a four-year sunset to sections 206, 207, 214, 215, 218, and 223. Section 504 authorizes consultation between FISA officers and law enforcement officers to coordinate efforts to investigate or protect against international terrorism, clandestine intelligence activities, or other grave hostile acts of a foreign power or an agent of a foreign power. Section 1003 authorizes the interception of electronic communications of computer trespassers under FISA.

5. 50 U.S.C. 1804(a)(7)(B) and 1823(a)(7)(B).

6. Senate Committee on the Judiciary, *Foreign Intelligence Surveillance Act of 1977*, S. Rept. 604, 95 Cong. 1 sess. (Government Printing Office [GPO], 1977), p. 7.

7. Ibid., p. 8, quoting the Senate Select Committee to Study Government Operations, *Intelligence Activities and the Rights of Americans*, S. Rept. 755, 94 Cong. 2 sess. (GPO, 1976), book 2, p. 12 (hereafter referred to as the "Church Committee Report").

8. Ibid. The Church Committee found and the 1977 Senate Judiciary Committee report on FISA acknowledged that "the infinite elasticity of the 'national security' criteria unrestrained by any judicial or external check, has been dramatically underscored in recent years by a series of surveillances directed against government employees and journalists for the avowed purpose of identifying the sources of 'leaks' of classified information." Senate Committee on the Judiciary, *Foreign Intelligence Surveillance Act of 1977*, p. 15, n. 27, quoting the Church Committee Report, book III, p. 321. See also Senate Select Committee on Intelligence, *Foreign Intelligence Surveillance Act of 1978*, S. Rept. 701, 95 Cong. 2 sess. (GPO, 1978), p. 42; House Permanent Select Committee on Intelligence, *Foreign Intelligence Surveillance Act of 1978*, H. Rept. 1283, 95 Cong., 2 sess. (GPO, 1978), pt. 1.

9. 407 U.S. 297, 300 (1972). Emphasis added.

10. The Court did not address and expressed no opinion as to "the issues which may be involved with respect to activities of foreign powers or their agents" (ibid, pp. 321–22) but recognized the possibility that departures from conventional law enforcement procedures for electronic surveillance (Title III) or physical search may be justified in domestic security cases and in cases involving foreign powers and agents (ibid., pp. 296–97). ("Moreover, we do not hold that the same standards and procedures prescribed by Title III are necessarily applicable to this case. We recognize that domestic security surveillance may involve different policy and practical considerations from the surveillance of 'ordinary crime.' . . . It may be that Congress, for example, would judge that the application and affidavit showing probable cause need not follow the exact requirements

of [Title III] but should allege other circumstances more appropriate to domestic security cases; that the request for prior court authorization could, in sensitive cases, be made to any member of a specially designated court [e.g., the District Court for the District of Columbia or the Court of Appeals for the District of Columbia Circuit]; and that the time and reporting requirements need not be so strict as those in [Title III].") Ibid.

11. House Permanent Select Committee, *Foreign Intelligence Surveillance Act of 1978*, p. 24.

12. The attorney general is required to "fully inform" the Intelligence Committees and submit reports on the total numbers of FISA applications and orders to the Judiciary Committees of both Houses of Congress. 50 U.S.C. 1863.

13. 50 U.S.C. 1808(a)(1).

14. Ibid. See *FBI Oversight in the 107th Congress by the Senate Judiciary Committee: FISA Implementation Failures,* An interim report by Senators Patrick Leahy, Charles Grassley, and Arlen Specter, February 2003, pp. 10–11 (expressing "disappoint[ment] with the nonresponsiveness of the DOJ and FBI" and "hope that the FBI and DOJ will reconsider their approach to congressional oversight in the future"). The Justice Department initially declined to provide classified answers to multiple questions about implementation and use of the FISA to the House Judiciary Committee in July 2002, prompting the Republican chairman of the House Judiciary Committee to threaten to obtain a subpoena if the information he requested from the Justice Department was not forthcoming. Steve Schultze, "Sensenbrenner Wants Answers on Act," *Milwaukee Journal Sentinel* (www.jsonline.com/news/nat/aug01/67685.asp [August 19, 2002]); Editorial, "Ashcroft—Above the Law?" *San Francisco Chronicle*, August 23, 2002, p. A38; Editorial, "Our Country, Our Freedom," *St. Louis Post-Dispatch*, September 10, 2002, p. B6.

15. *Congressional Record*, daily ed., October 25, 2001, pp. S10991–92. (Statement of Senator Leahy, on passage of the USA PATRIOT Act: "The sunset provision included in the final bill calls for vigilant legislative oversight, so that the Congress will know how these legal authorities are used and whether they are abuses over the next four years. . . . I agree with Leader Armey that the sunset will help ensure that law enforcement is responsive to congressional oversight and inquiries on use of these new authorities and that a full record is developed on their efficacy and necessity.")

16. 50 U.S.C. 1809(a)(7)(B); 1823(a)(7)(B).

17. Senate Committee on the Judiciary, *Foreign Intelligence Surveillance Act of 1977*, p. 53. An early version of the bill, S. 1536, expressly stated in a provision describing "Use of Information" that FISA-derived information concerning United States persons "may be used . . . for the enforcement of the criminal law if its use outweighs the possible harm to the national security." See 50 U.S.C. §1806, sec. 106.

18. 50 U.S.C. 1806(a), (c).

19. 50 U.S.C. 1806(d).

20. House Permanent Select Committee, *Foreign Intelligence Surveillance Act of 1978*, p. 41.

21. 50 U.S.C. 1806, 1825.

22. The minimization procedures allow for "the retention and dissemination of information that is evidence of a crime which has been, is being, or is about to be committed and that is to be retained or disseminated for law enforcement purposes." 50 U.S.C. 1806(h)(3).

23. Senate Select Committee on Intelligence, *Foreign Intelligence Surveillance Act of 1978*, pp. 41–42.

24. House Permanent Select Committee, *Foreign Intelligence Surveillance Act of 1978*, p. 39.

25. *In re All Matters Submitted to the Foreign Intelligence Surveillance Court*, 218 F. Supp. 2d 611, 620 (F.I.S.C. 2002).

26. A 1995 memorandum from the attorney general to the assistant attorney general, Criminal Division; the FBI director; the Counsel for Intelligence Policy; and the United States Attorneys, directs that logs be maintained of "all contacts with the Criminal Division, noting the time and participants involved in any contact, and briefly summarizing the content of any communication." See Janet Reno, "Memorandum: Procedures for Contacts between the FBI and the Criminal Division Concerning Foreign Intelligence and Foreign Counterintelligence Investigations," (July 19,1995), subsecs. A.4 and B.4.

27. *In re All Matters Submitted to the Foreign Intelligence Surveillance Court*, p. 620.

28. The Department of Justice brought a series of over seventy-five errors to the attention of the FISA Court in 2000 and 2001. Ibid.

29. Department of Justice, "Anti-Terrorism Act of 2001, Section-by-Section Analysis," September 19, 2001 (Consultation and Discussion Draft), p. 5.

30. Sylvia Adcock, Brian Donovan, and Craig Gordon, "America's Ordeal: Where the System Failed; Air Attack on Pentagon Indicates Weaknesses," *Newsday*, September 23, 2001, p. A03; Guy Gugliotta, "Terrorism 'Watch List' Was No Match for Hijackers," *Washington Post*, September 23, 2001, p. A22, David Willman and Alan C. Miller, "'Watch List' Didn't Get to Airline," *Los Angeles Times*, September 20, 2001, p. A1. See also Jeff Gerth, "CIA Chief Won't Name Officials Who Failed to Add Hijackers to Watch List," *New York Times*, May 15, 2003, p. A25.

31. Testimony of Associate Deputy Attorney General David S. Kris at *Protecting Constitutional Freedoms in the Face of Terrorism*, Hearing before the Senate Judiciary Subcommittee on the Constitution, Federalism, and Property Rights, 107 Cong. 1 sess. (GPO, October 3, 2001), p. 65.

32. Ibid., p. 60.

33. *Congressional Record*, daily ed., October 25, 2001, p. S11003–4 (statement of Senator Leahy).

34. Section 504 of the PATRIOT Act, titled "Coordination with Law Enforcement," amended 50 U.S.C. 1806(k) and 1825(k).

35. This proposal was discussed during consideration of the PATRIOT Act at no less than three hearings: *S.1448, the Intelligence to Prevent Terrorism Act of 2001 and Other Legislative Proposals in the Wake of the September 11, 2001 Attacks,* Hearing before the Senate Select Committee on Intelligence, 107 Cong.

1 sess. (GPO, September 24, 2001); *Homeland Defense,* Hearing before the Senate Judiciary Committee, 107 Cong. 1 sess. (GPO, September 25, 2001); *Protecting Constitutional Freedoms in the Face of Terrorism,* Hearing before the Senate Judiciary Subcommittee on the Constitution, Federalism, and Property Rights, 107 Cong. 1 sess. (GPO, October 3, 2001).

36. *Intelligence to Prevent Terrorism,* pp. 21–22, 29.

37. *Homeland Defense,* p. 25.

38. Ibid.

39. *Protecting Constitutional Freedoms,* pp. 82–89. One witness observed at this hearing that the Department of Justice "is no longer prepared to defend the constitutionality of its original proposal which it asked Congress to pass in five days" (p. 60, testimony of Mort Halperin, Senior Fellow, Council on Foreign Relations, and Chair, Advisory Board, Center for National Security Studies). In addition, on October 11, 2001, Senator Leahy stated, "Indeed, the Justice Department's own constitutional analysis provided to the committee at the request of our Members does not even attempt to justify the original proposal but instead presents an argument for why a change to 'a significant' purpose would be constitutional." *Congressional Record,* daily ed., October 11, 2001, p. S10593.

40. *In re Sealed Case No. 02-001,* 310 F.3d 717, 735–36 (F.I.S. Ct. Rev. 2002). Former Ashcroft Justice Department officials persist in arguing that the "significant purpose" change works to "eliminate the wall of separation between foreign threats and domestic crimes, and to allow law enforcement to be used as a weapon against terrorism," without noting the important limitation that the domestic crimes permitted to be investigated using FISA powers are those connected to foreign intelligence crimes only and not purely domestic terrorism. See Eric Posner and John Yoo, "The Patriot Act Under Fire," *Wall Street Journal,* December 9, 2003. John Yoo is a former Justice Department official under the Bush administration.

41. *In re Sealed Case No. 02-001,* p. 735.

42. Ibid. See also *ACLU* v. *U.S. Department of Justice,* 265 F. Supp. 2d 20, 31 (D.D.C. 2003).

43. *In re Sealed Case No. 02-001,* p. 735.

44. From the perspective of the intelligence community and the Senate Select Committee on Intelligence (SSCI), stumbling blocks to information sharing came from the law enforcement side of the wall, not the other way around. At one SSCI hearing, the point was made that "intelligence officers complain frequently that the Department of Justice refuses or is slow in sharing FISA information with intelligence agencies." *Intelligence to Prevent Terrorism,* p. 44.

45. 50 U.S.C. 403-5b(a)(1).

46. Ibid., 403–405(a)(2).

47. PATRIOT ACT, section 203.

48. Justin Rood, "FBI Wiretapping at 'Record' Levels, Thanks to Patriot Act; Bureau Can't Keep Up," *Congressional Quarterly Homeland Security,* February 24, 2004; see also Justin Rood, "FBI 'Drowning' in Information Harvested by Bugs and Wiretaps" (page15.com/2004/02/fbi-drowning-in-information-harvested.html [February 23, 2004]).

49. *In re Sealed Case No. 02-001,* p.741, n. 24.

50. The FISA Court of Review noted this issue and compared a defendant's right to obtain Title III applications and orders to challenge the legality of surveillance with a defendant's inability to review FISA applications and orders. The Court of Review stated, "Clearly, the decision whether to allow a defendant to obtain FISA materials is made by a district judge on a case by case basis, and the issue whether such a decision protects a defendant's constitutional rights in any given case is not before us." Ibid.

51. 18 U.S.C. App. 3.

52. See 50 U.S.C. 1806(f) (electronic surveillance), 1825(g) (physical searches), 1845(f)(2) (pen registers and trap and traces).

53. This fact was bluntly articulated by one court: "The language of section 1806(f) clearly anticipates that an ex parte, in camera determination is to be the rule. Disclosure and an adversary hearing are the exception, occurring *only* when necessary. . . . The government represents that it is unaware of any court ever ordering disclosure rather than conducting in camera and ex parte review, and the defendants cite no such case to the court. *United States* v. *Sattar*, No. 02 Cr. 395, 2003 U.S. Dist. LEXIS 16164 (S.D.N.Y., September 15, 2003).

54. See 18 U.S.C. 3500, and the following ones; 373 U.S. 83 (1963).

55. *United States* v. *Sattar*, p. 21; *United States* v. *Duggan*, 743 F.2d 59, 79 (2d Cir. 1984); *United States* v. *Ott*, 827 F.2d 473, 476 (9th Cir. 1987).

56. 50 U.S.C. 1806 (f); 50 U.S.C. 1825 (g).

57. *FBI Oversight in the 107th Congress*, p.19 and exhibit E.

58. The "lone wolf" provision contained in section 6001 of the Intelligence Reform and Terrorism Prevention Act amends FISA's definition of "agent of a foreign power" to cover a non-U.S. person who engages in international terrorism or activities in preparation for international terrorism, even absent any connection between the target and a foreign power. This amendment was adopted in 2004. See P.L. 108-458. Significant questions have been raised about the constitutionality of this provision. See *S.2586 and S.2659, Amendments to the Foreign Intelligence Surveillance Act*, Hearing before the Senate Select Committee on Intelligence, 107 Cong., 2 sess. (GPO, July 31, 2002), p. 46 (testimony of Jerry Berman, Executive Director, Center for Democracy and Technology).

59. Codified at 18 U.S.C. 2712.

60. 18 U.S.C. 2712(c).

61. "Before the department asks Congress for more powers, it needs to disclose how it is using the ones it already has. Yet the Justice Department has balked at reasonable oversight and public information requests." Editorial, "PATRIOT Act: The Sequel," *Washington Post*, February 12, 2003, p. A28. In addition, the American Bar Association in a resolution adopted February 10, 2003, urged the Congress, inter alia, to make available to the public "an annual statistical report on FISA investigations"; see www.epic.org/privacy/terrorism/fisa/aba_res_021003.html [November 2005].

62. P.L. 108-458, sec. 6002, "Additional Semiannual Reporting Requirements under the Foreign Intelligence Surveillance Act of 1978."

63. Statement of Senator Leahy on introduction of S. 436, Domestic Surveillance Oversight Act of 2003. *Congressional Record*, daily ed., February 25, 2003, p. S2705.

10

Why You Should Like the PATRIOT Act

JON KYL

American political history is littered with great issues and contro-versies that, while passionately contested in their time, seem little more than quaint eccentricities today. History has not been kind to—or has kindly forgotten—many critics of government actions that stirred heated opposition at the moment but seem common sense today. Do you remember the scenarios of doom painted by those who fought President Reagan's deregulation of gas prices in 1981 or, from your history books, recall how vigorously some opposed President Jefferson's Louisiana Purchase?

I predict that the USA PATRIOT Act debates will enter the same category in the years to come. The PATRIOT Act was a tailored and very necessary response to the events of September 11 and has shown itself an essential tool in the ongoing fight against terrorism. Indeed, it is something that this country should have done years ago. But truth does not automatically prevail as conventional wisdom, and even the seemingly obvious must be stated. With that thought in mind, I present my case for why the PATRIOT Act is a good and necessary law.

Why the PATRIOT Act Matters

Much of what the PATRIOT Act does seems technical and legalistic. Even in the accounts of its most vigorous defenders, the act does little more than fill gaps in the law. But these gaps matter. As we all know, in 2001 at least nineteen foreign terrorists were able to enter this country and plan and execute the most devastating terrorist attack this nation has suffered. The reasons why U.S. antiterror investigators failed to uncover and stop the September 11 conspiracy have now been explored by a Joint Inquiry of the House and Senate Intelligence Committees, other congressional committees and commissions, and the September 11 Commission. These commissions and inquiries have reviewed in painstaking detail the various pre–September 11 investigations that could have but did not prevent the September 11 plot. They show how close investigators came to discovering or disrupting the conspiracy, only to repeatedly reach dead ends or obstructions to their investigations.

In some of the most important pre–September 11 investigations, we now know exactly what stood in the way of a successful outcome: it was the laws that Congress wrote. Seemingly minor but nevertheless substantive gaps in our antiterror laws prevented the FBI from fully exploiting its best leads on the al Qaeda conspiracy.

One pre–September 11 investigation in particular came tantalizingly close to substantially disrupting or even stopping the terrorists' plot but was blocked by a flaw in our antiterror laws—one that has since been corrected by the PATRIOT Act. This investigation involved Khalid al-Mihdhar. Mihdhar was one of the eventual suicide hijackers of American Airlines Flight 77, which was crashed into the Pentagon, killing 58 passengers and crew and 125 people on the ground.

An account of a pre–September 11 investigation of Mihdhar is provided in the September 11 Commission's "Staff Statement No. 10." That statement notes (with some paraphrasing added) that

> during the summer of 2001 [a CIA agent] asked an FBI official . . . to review all of the materials [from an al Qaeda meeting in Kuala Lumpur, Malaysia] one more time. . . . [The FBI official] began her work on July 24 [of 2001]. That day she found the cable reporting that [Khalid al-]Mihdhar had a visa to the United States. A week later she found the cable reporting that Mihdhar's visa application—what was later discovered to be his first application—listed

New York as his destination. . . . [The FBI official] grasped the significance of this information.

[The FBI official] and an FBI analyst working the case promptly . . . met with an INS representative at FBI headquarters. On August 22 INS told them that Mihdhar had entered the United States on January 15, 2000, and again on July 4, 2001. . . . [The FBI agents] decided that if Mihdhar was in the United States, he should be found.[1]

At this point, the investigation of Khalid al-Mihdhar came up against the infamous legal "wall" that separated criminal and intelligence investigations at the time. The Joint Inquiry Report of the House and Senate Intelligence Committees describes what happened next:

> Even in late August 2001, when the CIA told the FBI, State, INS, and Customs that Khalid al-Mihdhar, Nawaf al-Hazmi, and two other "Bin Laden–related individuals" were in the United States, FBI headquarters refused to accede to the New York field office recommendation that a criminal investigation be opened, which might allow greater resources to be dedicated to the search for the future hijackers. . . . FBI attorneys took the position that criminal investigators "CAN NOT" (emphasis original) be involved and that criminal information discovered in the intelligence case would be "passed over the wall" according to proper procedures. An agent in the FBI's New York field office responded by e-mail, saying: "Whatever has happened to this, someday someone will die and, wall or not, the public will not understand why we were not more effective in throwing every resource we had at certain problems."[2]

The September 11 Commission has reached the following conclusion about the effect that the legal wall between criminal and intelligence investigations had on the pre–September 11 investigation of Khalid al-Mihdhar:

> Many witnesses have suggested that even if Mihdhar had been found, there was nothing the agents could have done except follow him onto the planes. We believe this is incorrect. Both Hazmi and Mihdhar could have been held for immigration violations or as material witnesses in the *Cole* bombing case. Investigation or interrogation of these individuals, and their travel and financial activities, also may have yielded evidence of connections to other

participants in the 9/11 plot. In any case, the opportunity did not arise.[3]

The PATRIOT Act dismantled the legal wall between intelligence and criminal investigations. However, it was enacted too late to have aided pre–September 11 investigations.

Another key pre–September 11 investigation also was blocked by a seemingly minor gap in the law. This case involves Minneapolis FBI agents' summer 2001 investigation of al Qaeda member Zacarias Moussaoui. Recent hearings before the September 11 Commission have raised agonizing questions about the FBI's pursuit of Moussaoui. Commissioner Richard Ben-Veniste noted the possibility that the Moussaoui investigation could have allowed the United States to possibly disrupt the September 11 plot. Commissioner Bob Kerrey even suggested that with better use of the information gleaned from Moussaoui, the "conspiracy would have been rolled up."[4]

Moussaoui was arrested by Minneapolis FBI agents several weeks before the September 11 attacks. That summer instructors at a Minnesota flight school became suspicious when Moussaoui, with little apparent knowledge of flying, asked to be taught to pilot a 747. The instructors contacted the Minneapolis office of the FBI, which immediately suspected that Moussaoui might be a terrorist.

FBI agents opened an investigation of Moussaoui and sought a Foreign Intelligence Surveillance Act (FISA) national security warrant to search his belongings. For three long weeks, these FBI agents were denied that FISA warrant. No search occurred before September 11. After the attacks (and largely because of them), the agents were issued an ordinary criminal warrant to search Moussaoui. Information in Moussaoui's belongings then linked him to two of the actual September 11 hijackers and to a high-level organizer of the attacks who later was arrested in Pakistan.

The September 11 commissioners are right to ask whether more could have been done to pursue this case, which was one of our best chances at stopping or disrupting the September 11 attacks. The problem is that given the state of the law at the time, the answer to the commissioners' question is probably "no." FBI agents were blocked from searching Moussaoui in part because of an outdated requirement of the 1978 FISA statute: that the target of an investigation be an agent of a "foreign power," such as a foreign government or terrorist group.[5] This law does not allow searches of apparent lone-wolf terrorists—suspects who have no known connection to any terror group. Senator Charles Schumer and I introduced legislation that would fill this gap in the law and allow the FBI to use FISA to monitor and search actual or apparent lone-wolf terrorists. That bill was unanimously approved by the Senate Judiciary Committee in April 2003 and was overwhelmingly approved by the full Senate that May. It was eventually included in the Intelligence Reform and Terrorism Prevention Act of 2004.[6]

What the various reports and commissions investigating the September 11 attacks have shown us so far is that where our antiterror laws are concerned, even seemingly little things can make a big difference. Before September 11, few would have thought that the lack of authority in FISA for the FBI to monitor and search lone-wolf terrorists might be decisive to our ability to stop a major terrorist attack on U.S. soil. And before September 11, though there was some attention to the problems posed by the perceived legal wall between intelligence and criminal investigations, and some efforts were made to lower that wall, there was little sense of urgency to the matter. These all seemed like legal technicalities—not problems that could eventually lead to the deaths of almost 3,000 people.

The PATRIOT Act and Intelligence Sharing

As the account of the al-Mihdhar investigation suggests, perhaps the single most important reform implemented by the PATRIOT Act was to repeal the legal barriers to information sharing between intelligence and criminal investigators. Because this part of the PATRIOT Act is often praised but infrequently described in detail, I offer the following accounts of pre–PATRIOT Act barriers to information sharing—and of the investigative successes that the removal of those barriers has made possible.

The FISA Court of Review decision upholding the PATRIOT Act's authorization for information sharing, *In re Sealed Case*, describes the origins of the pre–PATRIOT Act barriers:

> Apparently to avoid running afoul of the primary purpose test used by some courts, the 1995 [Attorney General] Procedures ["Procedures for Contacts between the FBI and the Criminal Division Concerning Foreign Intelligence and Foreign Counterintelligence Investigations"] limited contacts between the FBI and the Criminal Division in cases where FISA surveillance or searches were being conducted by the FBI for foreign intelligence (FI) or foreign coun-

terintelligence (FCI) purposes. The procedures state that "the FBI and Criminal Division should ensure that advice intended to preserve the option of a criminal prosecution does not inadvertently result in either the fact or the appearance of the Criminal Division's directing or controlling the FI or FCI investigation toward law enforcement objectives." Although these procedures provided for significant information sharing and coordination between criminal and FI or FCI investigations, based at least in part on the "directing or controlling" language, they eventually came to be narrowly interpreted within the Department of Justice, and most particularly by [the Justice Department's Office of Intelligence Policy and Review (OIPR)], as requiring OIPR to act as a "wall" to prevent the FBI intelligence officials from communicating with the Criminal Division regarding ongoing FI or FCI investigations. Thus the focus became the nature of the underlying investigation, rather than the general purpose of the surveillance. Once prosecution of the target was being considered, the procedures, as interpreted by OIPR in light of the case law, prevented the Criminal Division from providing any meaningful advice to the FBI.[7]

FBI director Robert Mueller, in testimony before the U.S. Senate Judiciary Committee in May 2004, provided a concrete account of the impact that these information-sharing barriers had on intelligence investigations:

Prior to September 11, an [FBI] agent investigating the intelligence side of a terrorism case was barred from discussing the case with an agent across the hall who was working the criminal side of that same investigation. For instance, if a court-ordered criminal wiretap turned up intelligence information, the criminal investigator could not share that information with the intelligence investigator—he could not even suggest that the intelligence investigator should seek a wiretap to collect the information for himself. If the criminal investigator served a grand jury subpoena to a suspect's bank, he could not divulge any information found in those bank records to the intelligence investigator. Instead, the intelligence investigator would have to issue a National Security Letter in order to procure that same information.[8]

Chicago U.S. attorney Patrick Fitzgerald, in an October 2003 hearing before the Judiciary Committee, described how these pre–PATRIOT Act information-sharing limits undercut one potentially vital terror investigation.

Mr. Fitzgerald discussed the grand jury testimony of Wadih el-Hage, a key member of the al Qaeda cell in Nairobi who, in September 1997, was apprehended while changing flights in New York City. Federal prosecutors subpoenaed el-Hage from the airport to testify before a federal grand jury in Manhattan. Mr. Fitzgerald described how el-Hage,

> provided some information of potential use to the intelligence community—including potential leads as to the location of his confederate Harun and the location of Harun's files in Kenya. Unfortunately, as el-Hage left the grand jury room, we knew that . . . [because of pre–PATRIOT Act restrictions] we would not be permitted to share the grand jury information with the intelligence community. . . . Fortunately, we found a way to address the problem that in most other cases would not work. Upon request, el-Hage voluntarily agreed to be debriefed by an FBI agent outside the grand jury room. . . . El-Hage then repeated the essence of what he told the grand jury to the FBI agent, including his purported leads on the location of Harun and his files. The FBI then lawfully shared the information with the intelligence community. In essence, we solved the problem by obtaining the consent of a since-convicted terrorist. We do not want to have to rely on the consent of al Qaeda terrorists to address the gaps in our national security.[9]

Mr. Fitzgerald went on to describe how, in August 1998, the American Embassy in Nairobi was bombed by al Qaeda. Investigators quickly learned that el-Hage's associate Harun was responsible. In this particular case, investigators had been able to work around information-sharing limits because of an al Qaeda terrorist's willingness to be interviewed by the FBI, and even with this information, U.S. agents were not able to stop a terrorist attack. The pre–PATRIOT Act limits were not a decisive factor in blocking U.S. intelligence agents from preventing the Kenya bombing—but they could have been. As U.S. Attorney Fitzgerald concluded, "We should not have to wait for people to die with no explanation [other] than that interpretations of the law blocked the sharing of specific information that probably [c]ould have saved [American lives]."[10]

As Attorney General Reno noted in her testimony before the September 11 Commission, "These restrictions [on information sharing] have now been eliminated as part of the PATRIOT Act."[11] Director Mueller, in his recent Judiciary Committee testimony, described the impact of this change:

The removal of the "wall" has allowed government investigators to share information freely. Now, criminal investigative information that contains foreign intelligence or counterintelligence, including grand jury and wiretap information, can be shared with intelligence officials. This increased ability to share information has disrupted terrorist operations in their early stages—such as the successful dismantling of the "Portland Seven" terror cell—and has led to numerous arrests, prosecutions, and convictions in terrorism cases.

In essence, before September 11 criminal and intelligence investigators were attempting to put together a complex jigsaw puzzle at separate tables. The PATRIOT Act has fundamentally changed the way we do business. Today, those investigators sit at the same table and work together on one team. They share leads. They fuse information. Instead of conducting parallel investigations, they are fully integrated into one joint investigation.[12]

These PATRIOT Act changes can directly be credited with some important recent successes in the war on terror. For example, in February 2003 federal prosecutors arrested and indicted Sami al-Arian and seven other suspected terrorists. The fifty-count indictment indicated that al-Arian was the financial director and the North American leader of Palestinian Islamic Jihad, a terrorist group that has killed more than 100 people in and around Israel, including two Americans. Al-Arian wired money to groups in Israel that paid money to the families of terrorists who carried out suicide bombings. He also founded three organizations in Florida that, among other things, drafted final wills and testaments for suicide bombers.

Incredibly, through much of the 1990s, al-Arian was secretly watched by two different sets of U.S. investigators. The FBI had been conducting a criminal probe of al-Arian since 1995. Meanwhile, intelligence agents had monitored al-Arian since the late 1980s. Because of pre–PATRIOT Act restrictions, the two sets of investigators were not able to share information and were not aware of the full extent of each other's investigations. It was only after the FISA Court of Review upheld the information-sharing provisions of section 203 of the act in November 2002 that intelligence officials were able to show their files to prosecutors. Several months after this provision of the PATRIOT Act was upheld and made effective, prosecutors arrested and indicted al-Arian and put an end to his activities.

The PATRIOT Act and Libraries

Of course, the PATRIOT Act includes much more than just the three provisions that facilitate information sharing. Although I will not discuss all of those provisions in detail, one provision has been a particular focus of attacks on the PATRIOT Act.

Section 215 of the PATRIOT Act allows the FBI to seek an order for "the production of tangible things (including books, records, papers, documents, and other items) for an investigation to obtain foreign intelligence information." FISA defines "foreign intelligence" as information relating to foreign espionage, foreign sabotage, or international terrorism, or information respecting a foreign power that relates to U.S. national security or foreign policy. Thus section 215 cannot be used to investigate ordinary crimes or even domestic terrorism. And in every case, a section 215 order must be approved by a judge, who must certify that the order is presented in the proper format.

Although section 215 is basically a form of subpoena authority, like that allowed for numerous other types of investigation, it has been heavily targeted by PATRIOT Act critics. Chief among their complaints is that section 215 could be used to obtain records from bookstores or libraries. Some of these critics have even alleged that section 215 would allow the FBI to investigate someone simply because of the books that he borrows from a library.

In fact, section 215 could be used to obtain library records, though neither this section nor any other provision of the PATRIOT Act specifically mentions libraries or is directed at libraries. Nevertheless, section 215 does authorize court orders to produce tangible records—which could include library records.

Where the critics are wrong is in suggesting that a section 215 order could be obtained *because* of the books that someone reads or the websites that he visits. Section 215 allows no such thing. Instead it allows an order to obtain "tangible things" as part of an investigation to "obtain foreign intelligence information"—information relating to foreign espionage or terrorism or relating to a foreign government or group and national security. As an added protection against abuse, the PATRIOT Act also requires that the FBI "fully inform" the House and Senate Intelligence Committees on all uses of section 215 every six months. These checks and safeguards leave FBI agents little room for the types of witch hunts that PATRIOT Act critics conjure up.

Furthermore, it bears mention that federal investigators already use an authority very similar to section 215—the grand jury subpoena—to obtain bookstore records. As Deputy Attorney General James Comey recently emphasized in a letter that he submitted to the editor of the *New York Times*, "Orders for records under [section 215] are more closely scrutinized and more difficult to obtain than ordinary grand jury subpoenas, which can require production of the very same records, but without judicial approval."[13] Similarly, in a September 2003 editorial, the *Washington Post* noted that investigative authority to review library records "existed prior to the PATRIOT Act; the law extends it to national security investigations, which isn't unreasonable."[14]

In addition, I would emphasize that an intelligence or criminal investigation may have good and legitimate reasons for extending to library or bookstore records. For example, in a recent domestic terrorism case, federal investigators sought to prove that a suspected bomber had built a particularly unusual detonator that had been used in several bombings. The investigators used a grand jury subpoena to show that the suspect had purchased a book giving instructions on how to build such a detonator.

Moreover, we should not forget that terrorists and spies historically *have* used libraries to plan and carry out activities that threaten U.S. national security. We know, for example, that some terrorists have used computers at public libraries to access the Internet and communicate by e-mail. It would be unwise to place libraries and bookstores beyond the scope of antiterror investigations.

Andrew McCarthy, a former federal prosecutor who led the 1995 terrorism case against Sheik Omar Abdel Rahman, recently elaborated on this point:

> Hard experience—won in the course of a string of terrorism trials since 1993—instructs us that it would be folly to preclude the government a priori from access to any broad categories of business record. Reading material, we now know, can be highly relevant in terrorism cases. People who build bombs tend to have books and pamphlets on bomb making. Terrorist leaders often possess literature announcing the animating principles of their organizations in a tone tailored to potential recruits. This type of evidence is a staple of virtually every terrorism investigation—both for what it suggests on its face and for the forensic significance of whose fingerprints may be on it. No one is convicted for having it—jurors are Americans,

too, and they'd not long stand for the odious notion that one should be imprisoned for the mere act of thinking.

When a defendant pleads "not guilty," however, he is saying: "I put the government to its proof on every element of the crime, including that I acted with criminal purport." Prosecutors must establish beyond a reasonable doubt not only that the terrorist engaged in acts but did so intending execrable consequences. If an accused says the precursor components he covertly amassed were for innocent use, is it not relevant that he has just borrowed a book that covers explosives manufacture? If he claims unfamiliarity with the tenets of violent jihad, should a jury be barred from learning that his paws have yellowed numerous publications on the subject? Such evidence was standard fare throughout Janet Reno's tenure as attorney general—and rightly so.[15]

Section 215 subpoenas should not be controversial. It is hardly anomalous that the FBI should be allowed to employ subpoenas under FISA. As the Justice Department's Office of Legal Policy recently noted in a published report, "Congress has granted some form of administrative subpoena authority to most federal agencies, with many agencies holding several such authorities."[16] The Justice Department "identified approximately 335 existing administrative subpoena authorities held by various executive branch entities under current law."[17]

Among the more frequently employed of existing executive subpoena authorities is permission for the attorney general to issue subpoenas "in any investigation of a federal health care offense."[18] As reported in the *Congressional Record*,

> According to the Public Law 106-544 Report, in the year 2001 the federal government used section 3486 to issue a total of 2,102 subpoenas in health care fraud investigations. These subpoenas uncovered evidence of "fraudulent claims and false statements such as 'upcoding,' which is billing for a higher level of service than that actually provided; double billing for the same visit; billing for services not rendered; and providing unnecessary services."[19]

One can hardly contend that although the federal government can use subpoenas to investigate Mohammed Atta if it suspects that he is committing Medicare fraud, it should not be allowed to use the same powers if it suspects that he is plotting to fly airplanes into buildings.

Finally, although the constitutionality of a tool so frequently used for so long might safely be assumed, it nevertheless merits describing exactly why subpoena power is consistent with the Fourth Amendment. A thorough explanation recently was provided by Judge Paul Niemeyer of the U.S. Court of Appeals for the Fourth Circuit. As Judge Niemeyer noted, the use of a subpoena does not require a showing of probable cause because a subpoena is not a warrant—it does not authorize an immediate physical intrusion of someone's premises in order to conduct a search. Rather, subpoenas are subject only to the Fourth Amendment's general reasonableness requirement—and they are reasonable in large part because of the continuous judicial oversight of their enforcement. As Judge Niemeyer stated in his opinion for the court in *In re Subpoena Duces Tecum*,

> While the Fourth Amendment protects people "against unreasonable searches and seizures," it imposes a probable cause requirement only on the issuance of warrants. U.S. Const. amend. IV ("and no warrants shall issue but upon probable cause, supported by oath or affirmation," etc.). Thus unless subpoenas are warrants, they are limited by the general reasonableness standard of the Fourth Amendment (protecting the people against "unreasonable searches and seizures"), not by the probable cause requirement.
>
> A warrant is a judicial authorization to a law enforcement officer to search or seize persons or things. To preserve advantages of speed and surprise, the order is issued without prior notice and is executed, often by force, with an unannounced and unanticipated physical intrusion. Because this intrusion is both an immediate and substantial invasion of privacy, a warrant may be issued only by a judicial officer upon a demonstration of probable cause—the safeguard required by the Fourth Amendment. *See* U.S. Const. amend. IV ("no Warrants shall issue, but upon probable cause"). The demonstration of probable cause to "a neutral judicial officer" places a "checkpoint between the government and the citizen" where there otherwise would be no judicial supervision.
>
> A subpoena, on the other hand, commences an adversary process during which the person served with the subpoena may challenge it in court before complying with its demands. As judicial process is afforded before any intrusion occurs, the proposed intrusion is regulated by, and its justification derives from, that process. . . . If [the appellant in this case] were correct in his assertion that investigative

subpoenas may be issued only upon probable cause, the result would be the virtual end to any investigatory efforts by governmental agencies, as well as grand juries. This is because the object of many such investigations—to determine whether probable cause exists to prosecute a violation—would become a condition precedent for undertaking the investigation. This unacceptable paradox was noted explicitly in the grand jury context in *United States* v. *R. Enterprises, Inc.*, where the Supreme Court stated: "The Government cannot be required to justify the issuance of a grand jury subpoena by presenting evidence sufficient to establish probable cause because the very purpose of requesting the information is to ascertain whether probable cause exists."[20]

The U.S. Supreme Court first upheld the constitutionality of subpoena authority in 1911. In *Wilson* v. *United States*, the Court concluded that "there is no unreasonable search and seizure when a writ, suitably specific and properly limited in scope, calls for the production of documents which . . . the party procuring [the writ's] issuance is entitled to have produced."[21]

The Supreme Court also has explicitly approved the use of subpoenas by executive agencies. In *Oklahoma Press Publishing Company* v. *Walling*, the Court found that the investigative role of an executive official in issuing a subpoena "is essentially the same as the grand jury's, or the court's in issuing other pretrial orders for the discovery of evidence."[22] Nearly fifty years ago, the Supreme Court in *Walling* was able to conclude that Fourth Amendment objections to the use of subpoenas by executive agencies merely "[raise] the ghost of controversy long since settled adversely to [that] claim."[23]

The PATRIOT Act and Delayed-Notice Searches

Another provision of the PATRIOT Act that has been the subject of much heat but little light is the authority for so-called "sneak and peak" searches. Section 213 of the act codifies judicial common law allowing investigators to delay giving notice to the target of a search that a search warrant has been executed against his property. Section 213 allows delayed notice of a search for evidence of any federal criminal offense if a federal court finds reasonable cause to believe that immediate notice may result in endangering the life or physical safety of an individual, flight from prosecution, destruction or tampering with evidence, intimidation of potential witnesses, or would otherwise seriously jeopardize the investigation. Notice still must be provided "within a reasonable period" of the warrant's execution, though this period may be extended by the court for good cause.

The ACLU, in particular, has been harshly critical of section 213, alleging that it "expands the government's ability to search private property without notice to the owner."[24] It has also stated that section 213 "mark[s] a sea change in the way search warrants are executed in the United States."[25] Finally, the ACLU has charged that as a result of section 213's authorization of delayed notice, "you may never know what the government has done."[26]

Not one of the allegations is remotely true. First, the target of a delayed-notice search will always eventually "know what the government has done" because section 213 expressly requires that the government give the target notice of the execution of the warrant "within a reasonable period of its execution." Section 213 clearly and explicitly authorizes only delayed notice, not no notice.

Furthermore, section 213 neither "expands the government's ability" to delay notice nor can section 213 even remotely be described as a "sea change" in the law. More than twenty-five years ago, the Supreme Court established that "covert entries are constitutional in some circumstances, at least if they are made pursuant to a warrant."[27] Congress first authorized delayed-notice searches in the 1968 Omnibus Crime Control and Safe Streets Act.[28] These searches repeatedly have been upheld as constitutional. For example, in 1990 the U.S. Court of Appeals for the Second Circuit held that

> certain types of searches or surveillance depend for their success on the absence of premature disclosure. The use of a wiretap, or a "bug," or a pen register, or a video camera would likely produce little evidence of wrongdoing if the wrongdoers knew in advance that their conversation or actions would be monitored. When nondisclosure of the authorized search is essential to its success, neither Rule 41 nor the Fourth Amendment prohibits covert entry.[29]

Similarly, the U.S. Court of Appeals for the Fourth Circuit has held that "the failure of the team executing the warrant to leave either a copy of the warrant or a receipt for the items taken did not render the search unreasonable under the Fourth Amendment. The Fourth Amendment does not mention notice, and the Supreme Court has stated that the Constitution does not categorically proscribe covert entries, which necessarily involve a delay in notice."[30]

To the extent that the ACLU intends to suggest that delayed-notice searches are unconstitutional, it bears mention that the Supreme Court has already addressed that view. In the 1979 *Dalia* case, the Supreme Court described the argument as "frivolous."[31]

Delayed-notice searches are important to antiterrorism investigations. The most common benefit of a delayed-notice warrant is that it allows investigators to uncover specific information about a terror suspect's activities or associates without tipping him off to the fact that he is under investigation. An inability to delay notice would, in many cases, seriously undermine the investigation. Some types of searches simply could not be conducted if notice could not be delayed—for example, there would be no point in conducting a wiretap if federal investigators were required to immediately inform the target that his conversations were being monitored. If immediate notice were required after every kind of search, terror suspects might flee the country, destroy computer files, alert associates of the investigation or stop communicating with them, injure or kill witnesses, or simply accelerate a planned attack.

According to the Justice Department, as of April 2003, it had obtained delayed-notice search warrants from courts forty-seven times and had been issued warrants allowing delayed notice of seizure fourteen times.[32] The most common period of delay of notice authorized under section 213 has been seven days. The courts have authorized specific periods of delay as short as one day and as long as ninety days, and have permitted delays of unspecified length that last until the indictment is unsealed.

In a number of recent cases, section 213 has played a key role in the success of an investigation. For example:

—In *United States* v. *Odeh*, a narcoterrorism case, a court issued a delayed-notice warrant pursuant to section 213 to allow the government to search an envelope that had been mailed to the target. This search confirmed that the target was operating a *hawala* money exchange that funneled money to the Middle East, including to an individual associated with Islamic Jihad in Israel. The delayed notice allowed investigators to conduct the search while continuing the investigation.[33]

—In a recent terrorism investigation, a terror suspect gave a closed box to another person who was cooperating with the FBI. The target had not authorized the FBI source to open the box and show the contents to others. Indeed, it appeared that the target had given the box to the source because the target was afraid that FBI agents would execute a warrant at his home and discover the items in the box. Delayed-notice authority allowed the FBI to obtain a warrant to search the contents of the box without immediately informing the target of the search. Immediate notification could have endangered the life of the source who was cooperating with the FBI.

—During the investigation of a domestic-terrorism group, agents followed one member of the group to what appeared to be a "safe house." After confirming that the house was used by the group, investigators obtained delayed-notice warrants to plant hidden microphones and cameras in the house. As a result, investigators discovered that the group was storing weapons and ammunition in the safe house. Another delayed-notice warrant was issued to allow the government to seize the weapons and ammunition. Several cell members were later arrested and convicted.

—During an investigation of a nationwide drug gang that distributes marijuana, cocaine, and methamphetamine, agents became aware that members were using a particular house. The members of this gang relied heavily on irregular use of cell phones and typically discontinued their use of telephones whenever a seizure of drugs occurred. Agents obtained a delayed-notice warrant and seized 225 kilograms of drugs at the house. Interceptions of phone conversations after the seizure revealed that the members of the gang thought that a rival gang had stolen their drugs. As a result, gang members continued to use their cell phones—thus allowing investigators to continue to monitor the gang.

Other PATRIOT Provisions

The importance to antiterror investigations of some of the remaining provisions of the PATRIOT Act recently was described by FBI Director Mueller in his testimony before the Senate Judiciary Committee. Director Mueller noted that

the PATRIOT Act gave federal judges the authority to issue search warrants that are valid outside the issuing judge's district in terrorism investigations. In the past, a court could only issue a search warrant for premises within the same judicial district—yet our investigations of terrorist networks often span multiple districts. The PATRIOT Act streamlined this process, making it possible for judges in districts where activities related to terrorism may have occurred to issue search warrants applicable outside their immediate districts.

In addition, the PATRIOT Act permits similar search warrants for electronic evidence such as e-mail. In the past, for example, if an agent in one district needed to obtain a search warrant for a subject's e-mail account, but the Internet service provider (ISP) was located in another district, he or she would have to contact an AUSA and agent in the second district, brief them on the details of the investigation, and ask them to appear before a judge to obtain a search warrant—simply because the ISP was physically based in another district. Thanks to the PATRIOT Act, this frustrating and time-consuming process can be averted without reducing judicial oversight. Today, a judge anywhere in the U.S. can issue a search warrant for a subject's e-mail, no matter where the ISP is based.

[Furthermore], the PATRIOT Act updated the law to match current technology, so that we no longer have to fight a twenty-first-century battle with antiquated weapons. Terrorists exploit modern technology such as the Internet and cell phones to conduct and conceal their activities. The PATRIOT Act leveled the playing field, allowing investigators to adapt to modern techniques. For example, the PATRIOT Act clarified our ability to use court-ordered pen registers and trap-and-trace devices to track Internet communications. The act also enabled us to seek court-approved roving wiretaps, which allow investigators to conduct electronic surveillance on a particular suspect, not a particular telephone—this allows them to continuously monitor subjects without having to return to the court.[34]

Civil Rights and Civil Liberties

In responding to some of the accusations of PATRIOT Act critics, I do not mean to dismiss the importance of either civil liberties or of independent oversight of the federal government. I would simply emphasize that the PATRIOT Act is carefully crafted legislation that both guarantees protection for civil liberties and is subject to ample oversight. I would note, in this vein, that in a report filed in March 2005, Department of Justice Inspector General Glenn A. Fine—an appointee of President Clinton—

described the results of his investigation of all recent civil rights and civil liberties complaints received by the Justice Department. Inspector General Fine found no incidents in which the PATRIOT Act was used to abuse civil rights or civil liberties.[35]

The PATRIOT Act's provisions for independent oversight of the new authorities created by the act were described in detail by Deputy Attorney General Comey in his April 2004 testimony before the Judiciary Committee. Mr. Comey noted:

> First, the USA PATRIOT Act preserves the historic role of courts by ensuring that the vital role of judicial oversight is not diminished. For example, the provision for delayed notice for search warrants requires judicial approval. In addition, under the act, investigators cannot obtain a FISA pen register unless they apply for and receive permission from federal court. The USA PATRIOT Act actually goes farther to protect privacy than the Constitution requires, as the Supreme Court has long held that law enforcement authorities are not constitutionally required to obtain court approval before installing a pen register. Furthermore, a court order is required to compel production of business records, in national security investigations.
>
> Second, the USA PATRIOT Act respects important congressional oversight by placing new reporting requirements on the department. Every six months, the attorney general is required to report to Congress the number of times section 215 has been utilized, as well as to inform Congress concerning all electronic surveillance under the Foreign Intelligence Surveillance Act. Under section 1001 of the USA PATRIOT Act, Congress receives a semiannual report from the department's inspector general detailing any abuses of civil rights and civil liberties by employees or officials of the Department of Justice. It is important to point out that in the inspector general's most recent report to Congress, he reported that his office has received no complaints alleging misconduct by department employees related to the use of a substantive provision of the USA PATRIOT Act.
>
> Finally, the USA PATRIOT Act fosters public oversight of the department. In addition to the role of the inspector general to review complaints alleging abuses of civil liberties and civil rights, the act provides a cause of action for individuals aggrieved by any

willful violation of Title III or certain sections of FISA. To date, no civil actions have been filed under this provision.[36]

A PATRIOT Act Interim Report Card

The United States has had some important successes in the war on terror so far. Worldwide, more than half of al Qaeda's senior leadership has been captured or killed. More than 3,000 al Qaeda operatives have been incapacitated. Within the United States, three different terrorist cells have been broken up—cells located in Buffalo, Seattle, and Portland. To date, 284 individuals have been criminally charged, and 149 individuals have been convicted or pleaded guilty, including shoe bomber Richard Reid, six members of the Buffalo terrorist cell, Ohio truck driver Iyman Faris, and U.S.-born Taliban John Walker Lindh.

PATRIOT Act–aided criminal prosecutions also have contributed to U.S. intelligence efforts to learn more about terrorist organizations. Facing long prison terms, some apprehended terrorists have chosen to cooperate with the U.S. government. So far, the Justice Department has obtained plea agreements from fifteen individuals who are now cooperating with terror investigations. One individual has given the United States information about weapons stored by terrorists in the United States. Another cooperating terrorist has given U.S. investigators information about locations in the United States that are being scouted or cased for potential attacks by al Qaeda.

The PATRIOT Act has played a major role in what U.S. antiterror investigations have accomplished so far. FBI Director Robert Mueller, for example, has not hesitated to give credit to the act. In his recent testimony before the Judiciary Committee, he also voiced strong support for renewing the law.

> For over two and a half years, the PATRIOT Act has proved extra-ordinarily beneficial in the war on terrorism and has changed the way the FBI does business. Many of our counterterrorism successes, in fact, are the direct results of provisions included in the act, a number of which are scheduled to "sunset" at the end of next year. I strongly believe it is vital to our national security to keep each of these provisions intact.[37]

Similarly, in an April 2004 field hearing before the Senate Judiciary Committee, Deputy Attorney General Comey stated that the PATRIOT Act "has made us immeasurably safer." He also responded to the allegation, occasionally made by some critics, that the PATRIOT Act was passed too quickly. He replied that "the USA PATRIOT Act was not rushed; it actually came ten years too late."[38]

The importance of the PATRIOT Act to American security also has drawn the attention of the September 11 Commission. Former New Jersey governor Thomas Kean has noted that the commission has had "witness after witness tell us that the PATRIOT Act has been very, very helpful, and if the PATRIOT Act, or portions of it, had been in place before September 11, that would have been very helpful."[39] Nor has such praise come only from the Republicans who have participated in the commission's proceedings. Former attorney general Janet Reno, for example, testified before the commission that "everything that's been done in the PATRIOT Act has been helpful."[40]

Such bipartisan expressions of support for the PATRIOT Act might help explain my confidence that the law not only deserves, but ultimately will receive, widespread acceptance as an appropriate and necessary investigative tool. The PATRIOT Act already has shown itself invaluable to antiterror investigations. If only it had already been in place in the summer of 2001, when FBI agents were tracking Khalid al-Mihdhar and Nawaf al-Hazmi.

I do not disparage those who would take on the role of the defender of civil liberties, suspicious of any new assertion of power. Building a free society is a never-completed task, always requiring vigilance against new threats to liberal institutions. The PATRIOT Act, however, is not a threat to free society. Rather, it is a key element to the defense of our free society.

Notes

1. National Commission on Terrorist Attacks upon the United States, "Threats and Responses in 2001. Staff Statement No. 10," *The 9/11 Commission Report* (www.9-11commission.gov/staff_statements/staff_statement_10.pdf [November 2005]).

2. Senate Select Committee on Intelligence and House Permanent Select Committee on Intelligence, *Joint Inquiry into Intelligence Community Activities before and after the Terrorist Attacks of September 11, 2001*, S. Rept. no. 107-351, H. Rept. no. 107-792, 107 Cong. 2d sess. (GPO, December 2002).

3. National Commission, "Staff Statement No. 10."

4. National Commission on Terrorist Attacks upon the United States, "Ninth Public Hearing," April 8, 2004 (www.9-11commission.gov/archive/hearing9/9-11Commission_Hearing_2004-04-08.pdf [December 2005]), p.54.

5. The Joint Inquiry and the September 11 Commission report note that other contributing factors to the failure to search Moussaoui include a misinterpretation of the FISA statute by FBI officials and the failure to connect Moussaoui to the Phoenix memo or to the entry of al-Mihdhar or al-Hazmi.

6. P.L. 108-408.

7. *In re Sealed Case No. 02-001*, 310 F.3d 717, 727–28 (F.I.S. Ct. Rev. 2002), citations omitted.

8. Senate Committee on the Judiciary, "Testimony of Robert S. Mueller, III, Director, Federal Bureau of Investigation," 108 Cong. 2 sess. (May 20, 2004).

9. Senate Committee on the Judiciary, *Protecting Our National Security from Terrorist Attacks: A Review of Criminal Terrorism Investigations and Prosecutions*, testimony of Patrick Fitzgerald, Unites States Attorney, Northern District of Illinois, 108 Cong. 1 sess. (October 21, 2003).

10. Ibid.

11. National Commission on Terrorist Attacks upon the United States, "Testimony of Janet Reno, Former Attorney General of the United States," April 13, 2004 (www.9-11commission.gov/hearings/hearing10/reno_statement.pdf [December 2005]).

12. Senate Committee on the Judiciary, "Testimony of Robert S. Mueller."

13. James Comey, "Rights and the Patriot Act," *New York Times*, April 28, 2004, p. A20.

14. Editorial, "PATRIOT (Act) Games," *Washington Post*, September 11, 2003, p. A22.

15. Andrew McCarthy, "PATRIOT Act under Siege," *National Review Online* (www.nationalreview.com/comment/mccarthy200311130835.asp [November 13, 2003]).

16. Department of Justice, Office of Legal Policy, *Report to Congress on the Use of Administrative Subpoena Authorities by Executive Branch Agencies and Entities, Pursuant to Public Law 106-544, Section 7* (May 13, 2002).

17. Ibid.

18. 18 U.S.C. 3486.

19. See *Congressional Record*, daily ed., June 22, 2004, vol. 150, pp. S7178–79. See also Office of Legal Policy, *Report to Congress*.

20. *In re Subpoena Duces Tecum*, 228 F.3d 341, 347–49 (4th Cir. 2000), citations omitted.

21. 221 U.S. 361 (1911).

22. 327 U.S. 186 (1946).

23. Ibid.

24. American Civil Liberties Union, "Surveillance under the USA PATRIOT Act," April 3, 2003 (www.aclu.org/safefree/general/17326res20030403.html [December 2005]).

25. ACLU statement October 23, 2001, as cited in Nat Hentoff, "Burglars with Badges: The Return of the Black Bag Job," *Village Voice*, December 3, 2001, p. 31.

26. From an ACLU advertisement, quoted in Ramesh Ponnoru, "1984 in 2003? Fears about the Patriot Act Are Misguided," *National Review*, June 2, 2003, pp. 17–18.

27. *Dalia* v. *United States*, 441 U.S. 238 (1979).

28. P.L. 90-351.

29. *United States* v. *Villegas*, 899 F.2d 1324, 1336 (2d Cir. 1990).

30. *United States* v. *Simons*, 206 F.3d 392, 403 (4th Cir. 2000).)

31. *Dalia* v. *United States*, 247.

32. Department of Justice, Office of Legislative Affairs, Letter from Jamie Brown, Acting Assistant Attorney General, to the House Committee on the Judiciary, May 13, 2003 (www.judiciary.house.gov/media/pdfs/patriotlet051303.pdf [December 2005]).

33. This and the other three examples given are from Department of Justice, Office of Legislative Affairs, Letter from William E. Moschella, Assistant Attorney General, to House Speaker Dennis Hastert, July 25, 2003 (www.cdt.org/security/usapatriot/030725doj.pdf [December 2005]).

34. Senate Committee on the Judiciary, "Testimony of Robert S. Mueller."

35. Department of Justice, Office of the Inspector General, *Report to Congress on Implementation of Section 1001 of the USA PATRIOT Act*, March 11, 2005.

36. Senate Committee on the Judiciary, *Preventing and Responding to Acts of Terrorism: A Review of Current Law*, statement of James B. Comey, Deputy Attorney General of the United States (Field Hearing, Salt Lake City, Utah, April 14, 2004).

37. Senate Committee on the Judiciary, "Testimony of Robert S. Mueller."

38. Senate Committee on the Judiciary, *Preventing and Responding to Acts of Terrorism*.

39. National Commission on Terrorist Attacks upon the United States, "Eighth Public Hearing," March 24, 2004 (www.9-11commission.gov/archive/hearing8/9-11Commission_Hearing_2004-03-24.pdf [December 2005]).

40. National Commission on Terrorist Attacks upon the United States, "Tenth Public Hearing," April 13, 2004 (www.9-11commission.gov/archive/hearing10/9-11Commission_Hearing_2004-04-13.pdf [December 2005]).

11

Why I Oppose the PATRIOT Act

RUSS FEINGOLD

M r. President, I have asked for this time to speak about the antiter-rorism bill before us, H.R. 3162. As we address this bill, we are especially mindful of the terrible events of September 11 and beyond, which led to the bill's proposal and its quick consideration in the Congress.

This has been a tragic time in our country. Before I discuss this bill, let me first pause to remember, through one small story, how September 11 has irrevocably changed so many lives. In a letter to the *Washington Post* recently, a man wrote that as he went jogging near the Pentagon, he came across the makeshift memorial built for those who lost their lives there. He slowed to a walk as he took in the sight before him—the red, white, and blue flowers covering the structure, and then, off to the side, a sec-

---

This text is excerpted from the statement made by Senator Feingold (D-Wisc.) from the Senate floor on October 25, 2001, during the debate over final passage of the PATRIOT Act. Senator Feingold was the only member of the Senate to vote against passage of the act. In the second part of this chapter, Clayton Northouse, editor of this volume, interviews Senator Feingold to get his take on the issues surrounding the passage and implementation of the PATRIOT Act.

ond, smaller memorial with a card. The card read, "Happy Birthday Mommy. Although you died and are no longer with me, I feel as if I still have you in my life. I think about you every day."

After reading the card, the man felt as if he were "drowning in the names of dead mothers, fathers, sons, and daughters." The author of this letter shared a moment in his own life that so many of us have had—the moment where televised pictures of the destruction are made painfully real to us. We read a card or see the anguished face of a grieving loved one, and we suddenly feel the enormity of what has happened to so many American families and to all of us as a people.

We all also had our own initial reactions, and my first and most powerful emotion was a solemn resolve to stop these terrorists. And that remains my principal reaction to these events. But I also quickly realized that two cautions were necessary, and I raised them on the Senate floor the day after the attacks.

The first caution was that we must continue to respect our Constitution and protect our civil liberties in the wake of the attacks. As the chairman of the Constitution Subcommittee of the Judiciary Committee, I recognize that this is a different world with different technologies, different issues, and different threats. Yet we must examine every item that is proposed in response to these events to be sure we are not rewarding these terrorists and weakening ourselves by giving up the cherished freedoms that they seek to destroy.

The second caution I issued was a warning against the mistreatment of Arab Americans, Muslim Americans, South Asians, or others in this country. Already, one day after the attacks, we were hearing news reports that misguided anger against people of these backgrounds had led to harassment, violence, and even death.

I suppose I was reacting instinctively to the unfolding events in the spirit of the Irish statesman John Philpot Curran, who said, "The condition upon which God hath given liberty to man is eternal vigilance."

During those first few hours after the attacks, I kept remembering a sentence from a case I had studied in law school. Not surprisingly, I didn't remember which case it was, who wrote the opinion, or what it was about, but I did remember these words: "While the Constitution protects against invasions of individual rights, it is not a suicide pact." I took these words as a challenge to my concerns about civil liberties at such a momentous time in our history; that we must be careful to not take civil liberties so literally that we allow ourselves to be destroyed.

But upon reviewing the case itself, *Kennedy* v. *Mendoza-Martinez,* I found that Justice Arthur Goldberg had made this statement but then ruled in favor of the civil liberties position in the case, which was about draft evasion. He elaborated:

> It is fundamental that the great powers of Congress to conduct war and to regulate the nation's foreign relations are subject to the constitutional requirements of due process. The imperative necessity for safeguarding these rights to procedural due process under the gravest of emergencies has existed throughout our constitutional history, for it is then, under the pressing exigencies of crisis, that there is the greatest temptation to dispense with fundamental constitutional guarantees which, it is feared, will inhibit governmental action.
>
> The Constitution of the United States is a law for rulers and people, equally in war and peace, and covers with the shield of its protection all classes of men, at all times, and under all circumstances. . . . In no other way can we transmit to posterity unimpaired the blessings of liberty, consecrated by the sacrifices of the revolution.[1]

I have approached the events of the past month and my role in proposing and reviewing legislation relating to it in this spirit. I believe we must redouble our vigilance. We must redouble our vigilance to ensure our security and to prevent further acts of terror. But we must also redouble our vigilance to preserve our values and the basic rights that make us who we are.

The Founders who wrote our Constitution and Bill of Rights exercised that vigilance even though they had recently fought and won the Revolutionary War. They did not live in comfortable and easy times of hypothetical enemies. They wrote a Constitution of limited powers and an explicit Bill of Rights to protect liberty in times of war as well as in times of peace.

There have been periods in our nation's history when civil liberties have taken a back seat to what appeared at the time to be the legitimate exigencies of war. Our national consciousness still bears the stain and the scars of those events: The Alien and Sedition Acts; the suspension of habeas corpus during the Civil War; the internment of Japanese Americans, German Americans, and Italian Americans during World War II; the blacklisting of supposed communist sympathizers during the McCarthy era; and the surveillance and harassment of antiwar protesters, including Dr. Martin Luther King Jr., during the Vietnam War. We must not allow these pieces of our past to become prologue. . . .

Now some may say, indeed we may hope, that we have come a long way since those days of infringements on civil liberties. But there is ample reason for concern. And I have been troubled in the past six weeks by the potential loss of commitment in the Congress and the country to traditional civil liberties.

As it seeks to combat terrorism, the Justice Department is making extraordinary use of its power to arrest and detain individuals, jailing hundreds of people on immigration violations and arresting more than a dozen "material witnesses" not charged with any crime. Although the government has used these authorities before, it has not done so on such a broad scale. Judging from government announcements, the government has not brought any criminal charges related to the attacks with regard to the overwhelming majority of these detainees.

For example, the FBI arrested as a material witness the San Antonio radiologist Albader al-Hazmi, who has a name like two of the hijackers and who tried to book a flight to San Diego for a medical conference. According to his lawyer, the government held al-Hazmi incommunicado after his arrest, and it took six days for lawyers to get access to him. After the FBI released him, his lawyer said, "This is a good lesson about how frail our processes are. It's how we treat people in difficult times like these that is the true test of the democracy and civil liberties that we brag so much about throughout the world." I agree with those statements. . . .

Even as America addresses the demanding security challenges before us, we must strive mightily also to guard our values and basic rights. We must guard against racism and ethnic discrimination against people of Arab and South Asian origin and those who are Muslim.

We who do not have Arabic names or do not wear turbans or headscarves may not feel the weight of these times as much as Americans from the Middle East and South Asia do. But as the great jurist Learned Hand said in a speech in New York's Central Park during World War II, "The spirit of liberty is the spirit which seeks to understand the minds of other men and women; the spirit of liberty is the spirit which weighs their interests alongside its own without bias . . . ." Was it not at least partially bias, however, when passengers on a Northwest Airlines flight in Minneapolis three weeks ago insisted that Northwest remove from the plane three Arab men who had cleared security?

Of course, given the enormous anxiety and fears generated by the events of September 11, it would not have been difficult to anticipate some of these reactions, both by our government and some of our people. Some have said rather cavalierly that in these difficult times, we must accept some reduction in our civil liberties in order to be secure.

Of course, there is no doubt that if we lived in a police state, it would be easier to catch terrorists. If we lived in a country that allowed the police to search your home at any time for any reason; if we lived in a country that allowed the government to open your mail, eavesdrop on your phone conversations, or intercept your e-mail communications; if we lived in a country that allowed the government to hold people in jail indefinitely based on what they write or think, or based on mere suspicion that they are up to no good, then the government would no doubt discover and arrest more terrorists.

But that probably would not be a country in which we would want to live. And that would not be a country for which we could, in good conscience, ask our young people to fight and die. In short, that would not be America.

Preserving our freedom is one of the main reasons that we are now engaged in this new war on terrorism. We will lose that war without firing a shot if we sacrifice the liberties of the American people. That is why I found the antiterrorism bill originally proposed by Attorney General Ashcroft and President Bush to be troubling.

The administration's proposed bill contained vast new powers for law enforcement, some seemingly drafted in haste and others that came from the FBI's wish list that Congress has rejected in the past. You may remember that the attorney general announced his intention to introduce a bill shortly after the September 11 attacks. He provided the text of the bill the following Wednesday and urged Congress to enact it by the end of the week. That was plainly impossible, but the pressure to move on this bill quickly, without deliberation and debate, has been relentless ever since.

It is one thing to shortcut the legislative process in order to get federal financial aid to the cities hit by terrorism. We did that, and no one complained that we moved too quickly. It is quite another to press for the enactment of sweeping new powers for law enforcement that directly affect the civil liberties of the American people without due deliberation by the peoples' elected representatives. Fortunately, cooler heads prevailed at least to some extent, and while this bill has been on a fast track, there has been time to make some changes and reach agreement on a bill that is less objectionable than the bill that the administration originally proposed.

As I will discuss in a moment, I have concluded that this bill still does not strike the right balance between empowering law enforcement and protecting civil liberties. But that does not mean that I oppose everything in the bill. Indeed, many of its provisions are entirely reasonable, and I hope they will help law enforcement more effectively counter the threat of terrorism.

For example, it is entirely appropriate that with a warrant the FBI be able to seize voice mail messages as well as tap a phone. It is also reasonable, even necessary, to update the federal criminal offense relating to possession and use of biological weapons. It made sense to make sure that phone conversations carried over cables would not have more protection from surveillance than conversations carried over phone lines. And it made sense to stiffen penalties and lengthen or eliminate statutes of limitation for certain terrorist crimes.

There are other noncontroversial provisions in the bill that I support: those to assist the victims of crime, to streamline the application process for public safety officers benefits and increase those benefits, to provide more funds to strengthen immigration controls at our northern borders, to expedite the hiring of translators at the FBI, and many others.

In the end, however, my focus on this bill, as chair of the Constitution Subcommittee of the Judiciary Committee in the Senate, was on those provisions that implicate our constitutional freedoms. And it was in reviewing those provisions that I came to feel that the administration's demand for haste was inappropriate; indeed, it was dangerous. Our process in the Senate, as truncated as it was, did lead to the elimination or significant rewriting of a number of audacious proposals that I and many other members found objectionable.

For example, the original administration proposal contained a provision that would have allowed the use in U.S. criminal proceedings against U.S. citizens of information obtained by foreign law enforcement agencies in wiretaps that would be illegal in this country. In other words, evidence obtained in an unconstitutional search overseas was to be allowed in a U.S. court.

Another provision would have broadened the criminal forfeiture laws to permit—prior to conviction—the freezing of assets entirely unrelated to an alleged crime. The Justice Department has wanted this authority for years, and Congress has never been willing to give it. For one thing, it touches on the right to counsel, since assets that are frozen cannot be used to pay a lawyer. The courts have almost uniformly rejected efforts to restrain assets before conviction unless they are assets gained in the alleged criminal enterprise. This proposal, in my view, was simply an effort on the part of the department to take advantage of the emergency situation and get something that they have wanted to get for a long time.

The foreign wiretap and criminal forfeiture provisions were dropped from the bill that we considered in the Senate. Other provisions were rewritten based on objections that I and others raised about them. For example, the original bill contained sweeping permission for the attorney general to get copies of educational records without a court order. The final bill requires a court order and a certification by the attorney general that he has reason to believe that the records contain information that is relevant to an investigation of terrorism.

So the bill before us is certainly improved from the bill that the administration sent to us on September 19 and wanted us to pass on September 21. But again, in my judgement, it does not strike the right balance between empowering law enforcement and protecting constitutional freedoms. Let me take a moment to discuss some of the shortcomings of the bill.

First, the bill contains some very significant changes in criminal procedure that will apply to every federal criminal investigation in this country, not just those involving terrorism. One provision would greatly expand the circumstances in which law enforcement agencies can search homes and offices without notifying the owner prior to the search. The longstanding practice under the Fourth Amendment of serving a warrant prior to executing a search could be easily avoided in virtually every case because the government would simply have to show that it has "reasonable cause to believe" that providing notice "may . . . seriously jeopardize an investigation." This is a significant infringement on personal liberty.

Notice is a key element of Fourth Amendment protections. It allows a person to point out mistakes in a warrant and to make sure that a search is limited to the terms of a warrant. Just think about the possibility of the police showing up at your door with a warrant to search your house. You look at the warrant and say, "Yes, that's my address, but the name on the warrant isn't me." And the police realize a mistake has been made and go away. If you're not home, and the police have received permission to do a "sneak and peek" search, they can come in your house, look around, and leave [without telling you for a "reasonable period"].

Another very troubling provision has to do with the effort to combat computer crime. The bill allows law enforcement to monitor a computer with the permission of its owner or operator, without the need to get a warrant or show probable cause. That is fine in the case of a so called "denial of service attack" or plain old computer hacking. A computer owner should be able to give the police permission to monitor communications coming from what amounts to a trespasser on the computer.

As drafted in the Senate bill, however, the provision might permit an employer to give permission to the police to monitor the e-mails of an employee who has [mis]used her computer at work to shop for Christmas gifts. Or someone who uses a computer at a library or at school and happens to go to a gambling or pornography site in violation of the Internet use policies of the library or the university might also be subjected to government surveillance—without probable cause and without any time limit. With this one provision, Fourth Amendment protections are potentially eliminated for a broad spectrum of electronic communications.

I am also very troubled by the broad expansion of government power under the Foreign Intelligence Surveillance Act, known as FISA. When Congress passed FISA in 1978, it granted to the executive branch the power to conduct surveillance in foreign intelligence investigations without meeting the rigorous probable cause standard under the Fourth Amendment that is required for criminal investigations. There is a lower threshold for obtaining a wiretap order from the FISA court because the FBI is not investigating a crime, it is investigating foreign intelligence activities. But the law currently requires that intelligence gathering be the primary purpose of the investigation in order for this lower standard to apply.

This bill changes that requirement. The government now will only have to show that intelligence is a "significant purpose" of the investigation. So even if the *primary* purpose is a criminal investigation, the heightened protections of the Fourth Amendment will not apply.

It seems obvious that with this lower standard, the FBI will try to use FISA as much as it can. And of course, with terrorism investigations that will not be difficult because the terrorists are apparently sponsored or at least supported by foreign governments. This means that the Fourth Amendment rights will be significantly curtailed in many investigations of terrorist acts.

The significance of the breakdown of the distinction between intelligence and criminal investigations becomes apparent when you see the other expansions of government power under FISA in this bill. One provision that troubles me a great deal is a provision that permits the government under FISA to compel the production of records from any business regarding any person, if that information is sought in connection with an investigation of terrorism or espionage.

Now we are not talking here about travel records pertaining to a terrorist suspect, which we all can see can be highly relevant to an investigation of a terrorist plot. FISA already gives the FBI the power to get airline, train, hotel, car rental, and other records of a suspect.

But under this bill, the government can compel the disclosure of the personal records of anyone—perhaps someone who worked with, or lived next door to, or went to school with, or sat on an airplane with, or has been seen in the company of, or whose phone number was called by the target of the investigation. And under this new provision, *all* business records can be compelled, including those containing sensitive personal information such as medical records from hospitals or doctors, or educational records, or records of what books someone has taken out of the library. This is an enormous expansion of authority under a law that provides only minimal judicial supervision.

Under this provision the government can apparently go on a fishing expedition and collect information on virtually anyone. All it has to allege in order to get an order for these records from the court is that the information is sought for an investigation of international terrorism or clandestine intelligence gathering. That is it. On that minimal showing in an ex parte application to a secret court, with no showing even that the information is *relevant* to the investigation, the government can lawfully compel a doctor or hospital to release medical records or a library to release circulation records. This is a truly breathtaking expansion of police power.

Let me turn to a final area of real concern about this legislation, which I think brings us full circle to the cautions I expressed on the day after the attacks. There are two very troubling provisions dealing with our immigration laws in this bill.

First, the administration's original proposal would have granted the attorney general extraordinary powers to detain immigrants indefinitely, including legal permanent residents. The attorney general could do so based on mere suspicion that the person is engaged in terrorism. I believe the administration was really overreaching here, and I am pleased that Senator Leahy was able to negotiate some protections. The Senate bill now requires the attorney general to charge the immigrant within seven days with a criminal offense or immigration violation. In the event that the attorney general does not charge the immigrant, the immigrant must be released.

While this protection is an improvement, the provision remains fundamentally flawed. Even with this seven-day charging requirement, the bill would nevertheless continue to permit the indefinite detention in two situations. First, immigrants who win their deportation cases could continue to be held if the attorney general continues to have suspicions. Second, this provision creates a deep unfairness to immigrants who are found not to be deportable for terrorism but have an immigration status violation, such as overstaying a visa. If the immigration judge finds that they are eligible for relief from deportation and therefore can stay in the country because, for example, they have longstanding family ties here, the attorney general could continue to hold them.

Now, I am pleased that the final version of the legislation includes a few improvements over the bill that passed the Senate. In particular, the bill would require the attorney general to review the detention decision every six months and would allow only the attorney general or deputy attorney general, not lower level officials, to make that determination. While I am pleased these provisions are included in the bill, I believe it still falls short of meeting even basic constitutional standards of due process and fairness. The bill continues to allow the attorney general to detain persons based on mere suspicion. Our system normally requires higher standards of proof for a deprivation of liberty. For example, deportation proceedings are subject to a clear and convincing evidence standard. Criminal convictions, of course, require proof beyond a reasonable doubt. The bill also continues to deny detained persons a trial or hearing where the government would be required to prove that the person is, in fact, engaged in terrorist activity. This is unjust and inconsistent with the values our system of justice holds dearly.

Another provision in the bill that deeply troubles me allows the detention and deportation of people engaging in innocent associational activity. It would allow for the detention and deportation of individuals who provide lawful assistance to groups that are not even designated by the secretary of state as terrorist organizations but instead have engaged in vaguely defined "terrorist activity" sometime in the past. To avoid deportation the immigrant is required to prove a negative: that he or she did not know, and should not have known, that the assistance would further terrorist activity.

This language creates a very real risk that truly innocent individuals could be deported for innocent associations with humanitarian or political groups that the government later chooses to regard as terrorist organizations. Groups that might fit this definition could include Operation Rescue, Greenpeace, and even the Northern Alliance fighting the Taliban in northern Afghanistan. This provision amounts to "guilt by association," which I believe violates the First Amendment. And speaking of the First Amendment, under this bill, a lawful permanent resident who makes a controversial speech that the government deems to be supportive of terrorism might be barred from returning to his or her family after taking a trip abroad. Despite assurances from the administration at various points in this process that these provisions that implicate associational activity would be improved, there have been no changes in the bill on these points since it passed the Senate.

Now here is where my cautions in the aftermath of the terrorist attacks and my concern over the reach of the antiterrorism bill come together. To the extent that the expansive new immigration powers that the bill grants to the attorney general are subject to abuse, who do we think is most likely to bear the brunt of that abuse? It won't be immigrants from Ireland, it won't be immigrants from El Salvador or Nicaragua, it won't even be immigrants from Haiti or Africa. It will be immigrants from Arab, Muslim, and South Asian countries. In the wake of these terrible events, our government has been given vast new powers, and they may fall most heavily on a minority of our population who already feel particularly acutely the pain of this disaster.

When concerns of this kind have been raised with the administration and supporters of this bill, they have told us, "Don't worry, the FBI would never do that." I call on the attorney general and the Justice Department to ensure that my fears are not borne out.

The antiterrorism bill that we consider in the Senate today highlights the march of technology and how that march cuts both for and against personal liberty. Justice Brandeis foresaw some of the future in a 1928 dissent, when he wrote:

> The progress of science in furnishing the government with means of espionage is not likely to stop with wiretapping. Ways may some day be developed by which the government, without removing papers from secret drawers, can reproduce them in court, and by which it will be enabled to expose to a jury the most intimate occur-

rences of the home. . . . Can it be that the Constitution affords no protection against such invasions of individual security?[2]

We must grant law enforcement the tools that it needs to stop this terrible threat. But we must give them only those extraordinary tools that they need and that relate specifically to the task at hand. In the play *A Man for All Seasons*, Sir Thomas More questions the bounder Roper whether he would level the forest of English laws to punish the devil. "What would you do?" More asks, "Cut a great road through the law to get after the Devil?" Roper affirms, "I'd cut down every law in England to do that." To which More replies,

> And when the last law was down, and the Devil turned round on you—where would you hide, Roper, the laws all being flat? This country's planted thick with laws from coast to coast . . . and if you cut them down . . . d'you really think you could stand upright in the winds that would blow then? Yes, I'd give the Devil benefit of law, for my own safety's sake.[3]

We must maintain our vigilance to preserve our laws and our basic rights. We in this body have a duty to analyze, to test, to weigh new laws that the zealous and often sincere advocates of security would suggest to us. This is what I have tried to do with this antiterrorism bill. And that is why I will vote against this bill when the roll is called.

Protecting the safety of the American people is a solemn duty of the Congress; we must work tirelessly to prevent more tragedies like the devastating attacks of September 11. We must prevent more children from losing their mothers, more wives from losing their husbands, and more firefighters from losing their heroic colleagues. But the Congress will fulfill its duty only when it protects *both* the American people and the freedoms at the foundation of American society. So let us preserve our heritage of basic rights. Let us practice as well as preach that liberty. And let us fight to maintain that freedom that we call America.

Interview with U.S. Senator Russ Feingold, April 21, 2004

CLAYTON NORTHOUSE: You were the only senator to vote against the PATRIOT Act. Why did you vote against it, and why didn't you have any support in doing so?

SENATOR RUSS FEINGOLD: The only way to explain this is to explain what it was like here after September 11. I had this very positive sense right after the attack that everybody was pulling together and that we were trying not to make mistakes. The reason that I thought I would be able to vote for such a bill was the experience I had with the first thing that we did right after September 11, which was the authorization of war allowing military action. The White House came out with a proposal in terms of authority to deal with September 11 issues that was in such vague language that it allowed the president to go after terrorism, essentially undefined. I remember seeing that put up in caucus and thinking, "Oh my goodness, this is going to be a big mistake if we don't tailor this to September 11 and the people that attacked us." Well, the leading Democrats in my caucus—people like John Kerry, Joe Biden, Carl Levin, and others—proposed an alternative, which we passed and which I voted for—all senators did—that specifically tailored the president's authority to September 11 and the groups that were involved in that attack. And so, I thought, this is a good sign.

A few days later came the USA PATRIOT Act—although it wasn't called that until the very end. The so-called antiterrorism law enforcement update legislation was starting to move, and we had some good feelings. We saw good things happening in the House, where Representatives Sensenbrenner and Conyers managed to get the entire committee from Bob Barr to Maxine Waters to vote for a version of the bill. Not all of it was acceptable, but basically, there was an attitude of, "There are some pretty extreme provisions; let's fix it." I was expecting the same process to occur in the Senate Judiciary Committee, where at the time I was the chairman of the Constitution subcommittee. I had every reason to believe that we would have a mark up in committee, but the bill was basically yanked out of committee by the Democratic leadership who, for whatever reason, decided to fold and go along with the White House demand that this thing be passed right away without amendment.

My view was—and I did look at the bill pretty carefully—90 percent of the actual provisions were reasonable, but the 10 percent that were not were a very serious threat to the Bill of Rights and the Constitution. So, my approach was to seek to offer amendments. I had reason to believe we would get somewhere. We had knocked off provisions in the early part of the process that were going to allow access to every kid's educational record. So, I thought, we should be able to get rid of these [other] things.

All of a sudden they wanted to bring the bill through the Senate with an agreement to have a couple hours of debate and no amendments, and I objected. I wouldn't let the bill go through. I had it out with the majority leader Tom Daschle, a friend of mine, and refused to let it go forward until I at least had a chance to offer a few of what I considered to be extremely reasonable amendments. And I got a few hours that night. I spoke briefly on each one, a few people said things about each amendment, and then the majority leader came out and said, "I don't want people to vote on the merits of these amendments. I want everybody to vote them down because we can't have any changes."

The rationale that some people used on my side of the aisle—what I considered to be folding on this—was that if we did open it up, it would get worse from a civil liberties point of view. Well, my view was that there were such serious problems with aspects of the bill that we needed to have a legitimate amendment process on something so fundamental. The process was shut down, the votes were taken very quickly, and I just felt like the bill needed to be changed in these respects to be acceptable. And so, I went up to the well that night—I think it was a Thursday night—and I saw that they had named it the USA PATRIOT Act. I took a good gulp and voted against it. I was not shocked to hear that others did not do so. The environment was supercharged emotionally at the time. It was very frightening.

NORTHOUSE: Let's go through the PATRIOT Act and outline some of the provisions you find troubling.

FEINGOLD: Among the ones that I really focused on in the debate about the bill itself and in subsequent discussions were the so-called computer trespass provisions, which pretty clearly allow the government to monitor somebody's use of a computer without any showing that the person was doing anything related to terrorism or even criminal conduct. It's triggered simply by a person doing something with a computer they're not supposed to do: if you buy a Christmas gift with your employer's computer, then you're a computer trespasser, and this may open you up. This provision does not target terrorism. This is an example of what Bob Novak was talking about—a few days after my vote when he said Russ Feingold was right—when he talked about an old wish list of the FBI.

A second provision is perhaps the most famous: section 215, the provision relating to business records—much of the discussion has been about library records. There is a myth out there the administration is perpetrating that this is a provision that involves real judicial review or discretion. It does not. The provision allows the government to come to a judge and say, "We seek this information in connection with a terrorist investigation." There is no requirement of any showing of evidence; there is no standard of "reasonable and articulable facts"—which is what the previous law required. The judge is required to approve the request without any showing of evidence. So, in fact, this is a very dangerous provision that is tied to neither terrorism nor criminal conduct because it relies simply on the say-so of the government without any judicial examination.

Similar problems exist with regard to the so-called sneak and peek provision, section 213. Again the question here is—and in none of these cases am I saying there shouldn't have been any change—should the government be able to go into somebody's house and be able to search with no notice to the people in the house for an indefinite period of time? The work I'm pursuing now with some of the Republican senators says, "You know, we're not saying there aren't situations where you need to do this, to do the sneak and peek search, but a judge should have to renew it every week. It shouldn't be indefinite." The thing that's troubling about this provision and the others that I've mentioned is the lack of either a requirement of nexus to terrorism or judicial review.

Roving wiretaps. I mentioned this on the night of the debate—it's one of the three amendments we chose to actually put to a vote. I'm not against a more elaborated and extensive roving wiretap authority. I don't have a problem with that. The president says, "If a terrorist is using one cell phone and a guy throws him another cell phone, shouldn't [law enforcement] be able to use the wiretap?" I don't disagree with that. What I think is that the government should have to ascertain and show in some way that there is reason to believe that this guy, the terrorist, is actually using these phones. This thing doesn't require that. It gives a broad authority, whether it be with phones or phone booths. Once they've got some connection to the phone booth, even if the phone booth is being used by completely innocent people, they've got complete access to the phone booth. So, that's an example of where I thought there's no reason to completely repeal it. You just have to require an ascertainment that the target is there.

NORTHOUSE: So now that we're not in the environment directly after the September 11 attacks, during the panicked moment during which the PATRIOT Act was passed, how are your suggestions of revisions being received?

FEINGOLD: First of all, I don't think we're in a different environment; I think we're in the same environment. We're in a very difficult, emotional time, and it's a scary time for a lot of people. So I am particularly impressed that the American people have risen up in opposition to the most disturbing parts of this legislation. Two hundred and seventy-five communities, I believe is the rough count, have passed resolutions either against the PATRIOT Act or saying that something ought to be done to fix it—including, incredibly, the New York City Council, which meets a good baseball throw away from the World Trade Center. Three or four state legislatures, typically the most conservative legislatures—Alaska, Idaho, Maine—have come out against it. I have never seen a response anything like this one to the USA PATRIOT Act.

Now, as to whether everybody knows what exactly is in this bill—clearly, they don't. But there is a sense out there that the government grabbed for power, and it has become so strong that some of the most conservative members of the Senate—Larry Craig, Mike Cappo, John Sununu (son of the former chief of staff for George H. W. Bush)—have created this bill for which I'm a key cosponsor, along with Dick Durbin. It's called the SAFE Act, and it is essentially about many of the things that we were just talking about.[4] So can you imagine how strong the sentiment must be against this bill that these senators—who are very loyal to the president, very loyal to his Iraq policy, and are about the least likely to question the president—have felt enormous political pressure to do something about it? And, in fact, it was a very conservative congressman from Idaho [C.L. "Butch" Otter] who passed an amendment in the House by a two-thirds margin on the sneak and peek provision.[5] So there is a powerful sentiment in this country that this needs to be fixed.

One of the most important things for me to do is to explain to people, "Look, ninety percent of the bill is not really controversial." The White House is distorting the debate by saying that we need the wall between the CIA and FBI broken down. Well, nobody disagrees with that. I didn't object to that provision. It's a bait and switch game. I'm talking about libraries and computers and law-abiding citizens. In fact, I remember saying at the time [of the vote on the PATRIOT Act], there are a lot of things in this bill I think we need to do: we should add more border guards [on the Canadian border]; if you wiretap somebody, you should be able to get their voicemail; we need more information sharing between the FBI and CIA. So that issue is a red herring; it's not what is in dispute.

NORTHOUSE: Do you think data mining has a useful purpose?

FEINGOLD: Yes. Data mining may help us to prevent acts of terrorism and crime. But while it is important to provide law enforcement the necessary tools to secure our safety, if left unchecked and unmonitored, data mining technologies could threaten the privacy and civil liberties of each and every American.

NORTHOUSE: How does data mining limit people's freedom?

FEINGOLD: In this day and age, when identity theft is becoming almost commonplace, lengthy searches at airports are routine, and faulty credit ratings have the power to ruin an individual, we need to be careful about how very personal information is gathered, maintained, and used. The idea of the government maintaining massive databases of personal information on citizens is inconsistent with the history of this country and the ideals of personal freedom and privacy that Americans have been raised to view as their birthright.

NORTHOUSE: What is the harm in having one's personal information stored in a database? If such databases prevent one terrorist attack, isn't this a harm that most people would be willing to endure?

FEINGOLD: The government hopes to be able to detect potential terrorists using data mining. Yet there is no evidence that data mining will, in fact, prevent terrorism. Data mining programs under development are being used to look into the future before being tested to determine if they would have been able to anticipate past events, like September 11 or the Oklahoma City bombing. Before we develop the ability to feed personal information about every man, woman, and child into a giant computer, we should learn what data mining can and can't do and what limits and protections for personal privacy are needed.

We must also consider the potential for errors in data mining. Most people don't even know what information is contained in their credit reports. Subjecting unchecked and uncorrected credit reports, medical records, and other documents to massive data mining makes the prospect of ensnaring innocents very real. If a credit agency has data about John R. Smith in John D. Smith's credit report, even the best data mining technology might reach the wrong conclusion and could even lead to the harassment or incarceration of innocent people.

NORTHOUSE: To the best of your knowledge, what government agencies are using data mining technologies?

FEINGOLD: I am aware of programs in the Department of Defense, the Department of Homeland Security, and the Department of Justice. One of the purposes of the Data Mining Reporting Act that I have introduced is to ensure that Congress is made aware of the specifics of all data mining programs that are currently in use or under development.[6] It would require governmental agencies to report to Congress about the existence of the various data mining programs now under way or being developed, and the impact those programs may have on our privacy and civil liberties so that Congress can determine whether the proposed benefits of a particular program come at too high a price.

The history is that Congress has learned about data mining programs only after millions of dollars have been spent and implementation is under way. The CAPPS II [Computer-Assisted Passenger Prescreening System] and MATRIX [Multistate Anti-Terrorism Information Exchange] programs illustrate this problem.[7] Both programs have laudable goals: we need an effective passenger screening system, and information sharing is important for law enforcement. But the failure to inform Congress and the public in a timely manner and to consider adequate protections has set back these programs. We need to have innovative and effective law enforcement programs, but we need to do it right.

NORTHOUSE: What does the government need to do to improve information sharing and prevent any failures similar to the failure to find the terrorists involved in the September 11 attack? Is there a way to do this without infringing on civil liberties?

FEINGOLD: Absolutely yes. As we have heard in testimony before Congress and before the September 11 Commission, information sharing does not mean sacrificing our civil liberties. Congress has an important role to play in making sure that the right people in law enforcement and the intelligence community are communicating with each other and that they have the most complete and accurate information available. Congress must continue to fund necessary technologies that will assist with information sharing, discard outmoded technology systems and ways of doing business that have led to stovepiping between the various agencies, and exercise strong oversight to ensure that personalities and bureaucratic turf battles don't get in the way of effective and appropriate law enforcement. At the same time, Congress can and should conduct oversight and provide for appropriate judicial review of law enforcement and terrorism prevention activities to protect privacy and civil liberties.

NORTHOUSE: Let's transition to the larger issue of balancing liberty and security in this environment. Many have argued that what's going on is a natural ebb and flow. During times of national crises, we pragmatically restrict liberty in order to more effectively protect the nation, and during times of peace and prosperity, we expand the protection of civil liberties. How do you respond to this type of argument, this pragmatism in the balance between liberty and security?

FEINGOLD: I wouldn't dispute that it's natural: so are death and taxes, but we try to limit them. It's natural that people respond this way at a time like this, but if there's any lesson from the history of this country, it's that serious mistakes have been made in times of crisis when the natural response is not checked. You can go back to the Alien and Sedition Acts, the suspension of habeas corpus during the Civil War, the measures that were taken against anti–war draft individuals in World War I, and most famous and desperately inappropriate, the internment of Japanese Americans during World War II. So we know from history that mistakes have been made, although we may find them understandable in a crisis, and we should be able to learn from those mistakes.

I thought things got off to a pretty good start almost within hours of September 11, when the president and others said that this should not be an occasion to single out Muslims and Asian-Americans and others in the way that sometimes happens during these times. That was a good message. Sadly, the USA PATRIOT Act was not a measured or necessary response. It was a power grab by the government, which saw an opportunity to take powers that were not tailored to the problem in the instances that I cited. So, in my view, it is not a simple question of a natural ebb and flow.

Look, I admit that when it comes to some of the things that we're talking about here, broadening powers in some of these provisions, I don't know if I would have looked at them exactly the same before September 11 as I do now. For example, on section 215, if somebody had said to me, "Gee, we need to get people's library records," I would have said, "Well, why do you need to get library records? What's so urgent?" I now recognize that if somebody in my community is known to have lunch with an al Qaeda person, I think we should be able to get that person's records. But a judge should say, yes, that's a good enough reason.

The administration's response was extreme because it knew it could get all it wanted because of the environment, and unfortunately, the administration took that opportunity. The administration didn't take the responsible step, and now it has a big problem on its hands. I would have liked to have voted for that bill. I consider the fight against terrorism the number one priority of this country, without a doubt. But the administration went too far and made the mistake of not taking the historic moment to get the balance right.

Notes

1. *Kennedy v. Mendoza-Martinez*, 372 U.S. 144 (1963).
2. *Olmstead v. United States*, 277 U.S. 438 (1928).
3. Robert Bolt, *A Man for All Seasons*, in *Three Plays* (London: Heinemann, 1967), act 1, p. 147.
4. *Security and Freedom Ensured Act of 2003 (SAFE Act)*, 108 Cong. 1 sess., S. 1709. The last major action taken on the SAFE Act was on October 2, 2003, when it was submitted to the Senate Judiciary Committee.
5. In July 2003 the House voted 309 to 118 to cut off funds for the implementation of section 213 of the PATRIOT Act, otherwise known as the sneak and peek provision. The Senate never passed this provision.

6. *Data Mining Reporting Act*, 108 Cong. 1 sess., S. 1544. The last major action taken on this bill was on July 31, 2003, when it was submitted to the Senate Judiciary Committee.

7. CAPPS II was a program of the Department of Homeland Security that was instituted to provide security at airports by assessing the risk level of passengers. CAPPS II searched through information stored in government and commercial databases and assigned a color-coded level of risk to each passenger. This program was discontinued in July 2004 and was succeeded by the Secure Flight program in November 2004, which included many of the same elements as the CAPPS II program.

MATRIX was a state-level program that combined criminal history records, driver's license data, vehicle registration records, and incarceration and corrections records with significant amounts of public data stored in commercial databases owned by a private company. MATRIX was discontinued on April 15, 2005.

Liberty and Security Timeline

This timeline offers an overview of some of the major events that have affected the relation between national security and civil liberties. A particular emphasis is given to the events following September 11, 2001.

1966
July 4   Freedom of Information Act (FOIA) is signed into law giving every member of the public the right of access to federal government records.
1967
December 18   In *Katz* v. *United States*, the Supreme Court finds that private conversations are protected by the Fourth Amendment.
1968
June 19   Omnibus Crime Control and Safe Streets Act is signed into law. This prohibits warrantless eavesdropping on telephone conversations, face-to-face conversations, and other forms of electronic communications and establishes procedures for receiving wiretapping warrants.
1974
November 21   FOIA is amended to incorporate judicial review of agency decisions, narrow some exemptions, restrict fees that agencies could charge, and impose a ten-day time limit on agencies to comply with a request. After these amendments, FOIA requests dramatically increase.
December 31   The most sweeping of all privacy legislation, the Privacy Act is signed into law protecting all federal government systems of records. It requires that

agencies store only relevant personal information, necessary for each agency's operations, and that the records stored by an agency be accurate and protected against unauthorized intrusion.

1976

April 26   Church Committee releases findings that expose the FBI's abuse of powers in the investigation of legitimate political activists and in keeping files on one million American citizens.

April 21   *United States* v. *Miller* rules that there is no expectation of privacy in information held by third parties. This finds that the Fourth Amendment is not applicable to data stored by companies in the private sector. Absent congressional legislation, the government does not need a search warrant to search or seize medical records, bank records, or anything else held by a third party.

May 18   Attorney General Edward Levi releases the Attorney General Guidelines on General Crimes, partly as a result of the Church Committee findings. The guidelines limit the investigative powers of the FBI, prohibit investigations of speech where there is no advocacy of violence, and mandate that there must be "specific and articulable" facts indicative of criminal activity in order to open an investigation.

1977

December 7 In *United States* v. *New York Telephone Co.*, the Supreme Court finds that the government's use of pen registers to record the telephone numbers an individual dials is not covered by the Fourth Amendment. In order to obtain approval for the installation of such registers, the government need only show that the information likely to be obtained is relevant to an ongoing criminal investigation.

1978

October 25 Foreign Intelligence Surveillance Act (FISA) is signed into law to regulate the collection of foreign intelligence information and limit surveillance to targeting foreign powers and their agents. FISA created the Foreign Intelligence Surveillance Court (FISA Court) and the Foreign Intelligence Surveillance Court of Review (FISA Court of Review) to grant or deny Justice Department requests for secret warrants.

1979

June 20   In *Smith* v. *Maryland*, the Supreme Court finds that the government's access to the data stored by "trap and trace" devices used to record dialed telephone numbers is not covered by the Fourth Amendment. The Court finds that the public has no reasonable expectation of privacy regarding the numbers one dials.

1980

October 15 Classified Information Procedures Act (CIPA) is signed into law. CIPA establishes procedures for limiting access to classified information in criminal cases while allowing defendants sufficient access to challenge the constitutionality and accuracy of the evidence.

1983

March 7   Attorney General William French Smith revises the Attorney General Guidelines on General Crimes by lowering the restriction on opening a criminal investigation from "specific and articulable" facts indicative of criminal activity to a "reasonable indication" of criminal activity. This is the standard currently in place.

1986

October 21   Electronic Communications Privacy Act (ECPA) is signed into law. ECPA establishes procedures for the government's access to and interception of electronic communications.

1988

August 12   The first public report is released on ECHELON. The National Security Agency in conjunction with the United Kingdom, Canada, Australia, and New Zealand operates the ECHELON program, which captures a great number of nondomestic communications and searches the vast amount of data for certain keywords.

1994

October 25   Communication Assistance for Law Enforcement Act (CALEA) requires telecommunications companies to develop and use technologies that are capable of being used for interception of communications by law enforcement, for instance, by enabling the use of wiretap devices.

2000

August 15   In *United States Telecom Association* v. *Federal Communications Commission*, the U.S. Court of Appeals for the District of Columbia finds that CALEA does not require carriers to disclose packet-mode data to the government absent court approval.

July 11   The FBI is reported to be using a technology called Carnivore (later changed to DCS 1000), which is installed at a node of an Internet service provider and monitors all information traveling through this node. The FBI states that Carnivore filters out and stores only that information relevant to an FBI investigation.

2001

September 11   Attacks on the World Trade Center and the Pentagon destroy the former and badly damage the latter. In the weeks after, 1,200 people are detained, 752 on immigration charges.

September 18   Authorization for Use of Military Force is passed by Congress and signed by President Bush. The resolution authorizes the president to use "all necessary and appropriate force against those nations, organizations, or persons he determines planned, authorized, committed, or aided the terrorist attacks" or "harbored such organizations or persons, in order to prevent any future acts of international terrorism against the United States by such nations, organizations, or persons."

September 18   Justice Department issues an interim regulation allowing for the detention of non–U.S. citizen terrorist suspects for up to forty-eight hours or "an additional reasonable period of time."

September 21   The nation's chief immigration judge, Judge Michael J. Creppy, issues instructions to hundreds of judges for immigration cases stating that "each case is to be heard separately from all other cases on the docket. The courtroom must be closed for these cases—no visitors, no family, and no press." The instructions also note that "the restriction includes confirming or denying whether such a case is on the docket."

October 26   After passing in the Senate 97 to 1 and in the House 337 to 79, the PATRIOT Act is signed into law. Among other things, the law provides new authority for wiretapping, monitoring Internet communications, and sharing information and the use of roving wiretaps and sneak-and-peek searches.

October 31   The Justice Department issues a regulation that permits, without a court order, surveillance of attorney-client conversations wherever there is "reasonable suspicion" that the inmate is using the conversation to facilitate acts of terrorism.

November 13   President Bush issues an executive order authorizing the creation of military tribunals to try noncitizens alleged to be international terrorists. The Department of Defense is to establish the rules and procedures of the tribunals.

November 19   Aviation and Transportation Security Act (ATSA) is signed into law and directs that all passengers be subject to a computer-assisted prescreening system. The act created the Transportation Security Administration (TSA).

2002

January   TSA's Office of National Risk Assessment is charged with developing the Computer Assisted Passenger Prescreening System (CAPPS II). The program is expected to increase security at airports by assessing the risk level of passengers. CAPPS II will search through information stored in government and commercial databases and assign a color-coded level of risk to each passenger.

April 10   The Department of Defense's Defense Advanced Research Projects Agency (DARPA) program, Total Information Awareness (TIA), is publicly revealed in testimony before the Senate Armed Services Committee. TIA's mission is to create "ultra-large all-source information repositories" that would be populated by data covering "communications, financial, education, travel, medical, veterinary, county entry, place/event history, transportation, housing, critical resources, and government." It is to create data-mining tools to search through this massive amount of data to discover patterns of activity relating to terrorism.

May 17   In *In Re All Matters Submitted to the Foreign Intelligence Surveillance Court*, the Foreign Intelligence Surveillance Act Court issues its opinion, criticizing the Department of Justice for giving the court misleading information in seventy-five cases. The court limits Justice's request to share intelligence information for criminal prosecutions and upholds the "wall" dividing criminal and intelligence operations of the Justice Department and FBI.

May 30   Attorney General John Ashcroft revises the Attorney General's Guidelines on General Crimes, Racketeering Enterprise, and Terrorism Enterprise Investigations, modifying the standards used for permissible searches by FBI agents. The standards expand the use of the Internet and libraries as well as the monitoring of religious organizations. Previously the FBI could not search for new leads using the Internet but could only use it in connection with an established investigation.

October 8   In *North New Jersey Media Group v. Ashcroft*, the U.S. Court of Appeals for the Third Circuit reverses the ruling of the Newark District Court, which ordered all deportation hearings to be open unless the government is able to show a need for a closed hearing on a case-by-case basis.

November 18   In *In re: Sealed Case No. 02-001* (F.I.S. Ct. Rev. 2002)—the first appeal of the FISA Court since the passage of FISA in 1978—the Foreign Intelligence Surveillance Court of Review overturns the FISA Court's ruling. The Court of Review finds that there need be no wall dividing the operations of the criminal and intelligence branches of the Department of Justice and that prosecutors may take an active role in deciding how to use wiretaps authorized by the FISA Court.

November 22   Homeland Security Act of 2002 is signed into law. The act creates the

	Department of Homeland Security, the largest reorganization of the government in more than fifty years.
December 17	E-Government Act is signed into law. The act requires federal government agencies to post privacy policies on their websites and to submit privacy impact assessments before acquiring or developing new information technologies.

2003

January 8	In *Hamdi v. Rumsfeld*, the U.S. Court of Appeals for the Fourth Circuit finds that Hamdi's military detention is permitted by the Authorization for Use of Military Force and that Hamdi cannot challenge his classification as an enemy combatant. Yasir Asem Hamdi is a U.S. citizen captured in a combat zone in Afghanistan and is held without charges and without access to counsel in a naval brig in Norfolk, Va.
February 24	An amendment proposed by Senator Ron Wyden is signed into law. The amendment prohibits funding for the Total Information Awareness program unless a report is given to Congress on the effectiveness of TIA, the laws applicable to it, and its effects on the public's right to privacy. The amendment also prohibits the use of American citizens' personal information without the explicit permission of Congress.
March 11	In *Al Odah v. United States*, the U.S. Court of Appeals for the District of Columbia finds that Al Odah, who was detained by United States forces in Afghanistan and is detained at Guantanamo Bay, Cuba, does not have access to U.S. courts and thereby cannot challenge his classification as an enemy combatant.
March 24	Supreme Court refuses to permit civil liberties groups to file for an appeal of the FISA Court of Review's ruling of November 18, 2002. The groups are not allowed to appeal because only the government was a party to the case, and therefore only the government can file for an appeal.
March	TSA begins developing the CAPPS II program.
May 1	A proposal is made in Congress to give the CIA and Department of Defense national security letter powers to require Internet service providers, telecommunications companies, and others to produce phone records, e-mail logs, and other information without court approval. Under the PATRIOT Act, this authority only rests with the FBI.
	The Department of Justice releases a report stating that the department used secret warrants pursuant to FISA a record 1,228 times—an increase of 30 percent over 2001—and that no requests for the warrant were turned down by the FISA Court.
	The Terrorism Threat Integration Center, originally proposed by President Bush in his 2003 State of the Union Address, begins operating. The center is to integrate all threat information from the CIA, FBI, Department of Homeland Security, and Department of Defense.
May 20	The mandated report regarding the TIA program is delivered to Congress. DARPA changes the name of Total Information Awareness to Terrorism Information Awareness.
June 17	In *Center for National Security Studies* v. *Department of Justice*, the U.S. Court of Appeals for the District of Columbia overturns the Federal District Court's ruling that ordered the release of the names of detainees held by the Department of Justice within fifteen days in compliance with FOIA. As of August 2, 2002 (the date of the district court's ruling), 81 of the 752 remain detained on immigration charges since September 11. The appeals

	court finds that the Department of Justice is not required to release the names because FOIA makes exceptions for "law enforcement records"—investigating materials not available in routine cases.
July	First reports are issued concerning the Multistate Anti-Terrorism Information Exchange (MATRIX) program, which combines criminal history records, driver's license data, vehicle registration records, and incarceration and corrections records with significant amounts of public data records stored in commercial databases owned by a private company. The information can be accessed by police officers working in the field. There are sixteen states participating in the program.
September 25	Department of Defense Appropriations Act is passed, and Section 8131 bans funding for the Terrorism Information Awareness program.
December 18	In *Padilla* v. *Rumsfeld*, the U.S. Court of Appeals finds that "Padilla's detention is not authorized by Congress, and, absent such authorization, the President does not have the power . . . to detain as an enemy combatant an American seized on American soil outside a zone of combat." Suspected of planning a dirty bomb attack, Jose Padilla is being held without charges and without access to counsel in a navy brig in Charleston, S.C. The court orders that Padilla should be released from military custody within thirty days and that the "government can transfer Padilla to appropriate civilian authorities who can bring criminal charges against him."
December 22	The Advisory Panel to Assess Domestic Response Capabilities for Terrorism Involving Weapons of Mass Destruction, created in 1998 to investigate the terrorist attacks in Tanzania and Kenya, releases a report that recommends a "civil liberties oversight board" with bipartisan membership to review how constitutional guarantees are being affected by new laws and regulations enhancing national security.

2004

January 23	In *Humanitarian Law Project* v. *Ashcroft*, a U.S. District Court ruled that the USA PATRIOT Act's ban on providing "expert advice and assistance" to terrorist groups is in violation of the First and Fourth Amendments to the Constitution because it "could be construed to include unequivocally pure speech and advocacy protected by the First Amendment."
March	State participants in the MATRIX program are down to five states: Connecticut, Florida, Michigan, Ohio, and Pennsylvania. Oregon, Utah, California, Texas, Louisiana, Alabama, Georgia, South Carolina, Kentucky, Wisconsin, and New York have dropped out of the program.
March 1	The Report of the Technology and Privacy Advisory Committee, "Safeguarding Privacy in the Fight Against Terrorism," is released. The report gives an analysis of Terrorism Information Awareness and the privacy and civil liberty issues surrounding data-mining activities of the government.
June 28	The Supreme Court issues three opinions fundamentally affecting the relation between national security and civil liberties: *Hamdi* v. *Rumsfeld*, *Rasul* v. *Bush*, and *Padilla* v. *Rumsfeld*. In *Hamdi* the Court vacates the U.S. Court of Appeals decision and finds that a U.S. citizen held in the territory of the United States as an enemy combatant must be afforded due process to challenge the classification as an enemy combatant and must have access to counsel. In *Rasul* and *Al Odah* v. *United States* (decided together), the Court ruled that Guantanamo is not beyond the jurisdiction of U.S. courts and that Rasul and Al Odah have the right to challenge their detention in U.S. courts. In *Padilla* the Court states that Padilla filed in the wrong court and

against the wrong defendants and that in order to challenge his detention, Padilla must refile in a lower court.

August CAPPS II is terminated.

September 29 In *John Doe* v. *Ashcroft*, the Manhattan U.S. District Court upholds a challenge to the national security letter (NSL) power in the PATRIOT Act, which gives the FBI the power to issue a subpoena calling for the release of personal information sought in a national security investigation without court review and bars the company from disclosing to a lawyer or anyone else that they have received a subpoena. The law was challenged by an Internet service provider who received an NSL from the FBI.

November The Department of Homeland Security creates a new program called Secure Flight, which is similar to the CAPPS II program. The Transportation Security Administration orders airlines to turn over millions of passenger records to test the program.

December 17 President Bush signs the Intelligence Reform and Terrorism Prevention Act, which brings together fifteen separate intelligence agencies under the Director of National Intelligence and creates the Privacy and Civil Liberties Oversight Board and a counterterrorism center. The act also includes the "lone wolf" amendment, which enables the government to receive FISA warrants for individuals who are not connected with a foreign power or terrorist organization.

2005

January It is reported that Carnivore (DCS 1000) is no longer used by the FBI; commercially available software is used instead.

April 15 Amid controversy and concern over privacy, the MATRIX program is discontinued.

June 29 President Bush issues an executive order creating a national security division within the FBI that falls under the direction of the director of national intelligence and is intended to merge foreign and domestic intelligence.

September 9 U.S. District Court in Bridgeport, Conn., rules in *John Doe* v. *Gonzales* that an NSL to a library consortium violated the right to free speech. This decision, along with the September 29, 2004, U.S. District Court decision, was stayed pending a November 2, 2005, appeal before the U.S. Court of Appeals for the Second Circuit.

November 6 In the *Washington Post*, Barton Gellman ("The FBI's Secret Scrutiny," p. A1) reports that the Justice Department has used the NSL power of section 505 of the PATRIOT Act (see entry for September 29, 2004) to secretly issue over 30,000 mandatory requests a year for information.

December 16 In the *New York Times*, James Risen and Eric Lichtblau ("Bush Lets U.S. Spy on Callers without Courts," p. A1) report that "President Bush secretly authorized the National Security Agency to eavesdrop on Americans and others inside the United States to search for evidence of terrorist activity without court-approved warrants." Circumventing the existing procedures of the FISA Court, the 2002 secret executive orders permitted the monitoring of international telephone calls and emails of "hundreds, perhaps thousands, of people inside the United States."

December As this volume goes to print, Congress extends expiring sections of the PATRIOT Act for one month. Congress intends to review renewal of these sunset provisions early in 2006. Under discussion will be permanently extending fourteen of the sixteen provisions. Four-year sunsets would be placed on section 215, which allows the government to access library and business records; section 206, which allows for roving wiretaps; and the "lone wolf" provision. A bipartisan group of senators had threatened to block the bill unless civil liberties safeguards were put in place.

2006

February   Senate Judiciary Committee holds a hearing on "Wartime Executive Power and the NSA's Surveillance Authority."

Source: Compiled by Clayton Northouse.

Further Resources

The following are resources dealing with national security, civil liberties, and information privacy, with a focus on recent materials. This is by no means a comprehensive bibliography for these topics but does offer some important next steps in covering the issues discussed in this volume.

Abdolian, Lisa Finnegan, and Harold Takooshian. 2003. "The USA Patriot Act: Civil Liberties, the Media, and Public Opinion." *Fordham Urban Law Journal* 30, no. 4: 1429–53.

Agre, Philip, and Marc Rotenberg, eds. 1998. *Technology and Privacy: The New Landscape*. MIT Press.

Alderman, Ellen, and Caroline Kennedy. 1995. *The Right to Privacy*. New York: Knopf.

American Civil Liberties Union. 2001. "USA Patriot Act Boosts Government Powers While Cutting Back on Traditional Checks and Balances." ACLU Legislative Analysis. New York (November).

Aronov, Rita F. 2004. "Privacy in a Public Setting: The Constitutionality of Street Surveillance." *Quinnipiac Law Review* 22, no. 4: 769–810.

Baker, Nancy V. 2003. "National Security versus Civil Liberties." *Presidential Studies Quarterly* 33, no. 3: 547–67.

Baldwin, Fletcher N. Jr. 2004. "The Rule of Law, Terrorism, and Countermeasures Including the USA Patriot Act of 2001." *Florida Journal of International Law* 16, no. 1: 43.

Ball, Howard. 2004. *The USA Patriot Act of 2001 : Balancing Civil Liberties and National Security: A Reference Handbook*. Santa Barbara, Calif.: ABC-CLIO.

Bandes, Susan. 2002. "Power, Privacy and Thermal Imaging." *Minnesota Law Review* 86, no. 6: 1379–91.

Bellia, Patricia L. 2004. "Surveillance Law through Cyberlaw's Lens." *George Washington Law Review* 72 (August): 1375.

Berkowitz, Bruce D. 1989. *Strategic Intelligence*. Princeton University Press.

_____. 2003. *The New Face of War*. Free Press.

Berkowitz, Bruce D., and Allan E. Goodman. 2000. *Best Truth: Intelligence in the Information Age*. Yale University Press.

Berman, Jerry. 2003. "Statement of Jerry Berman, president, Center for Democracy and Technology, before the House Committee on the Judiciary and the House Select Committee on Homeland Security" (www.cdt.org/testimony/030722berman.pdf [November 2005]).

Berman, Jerry, and Lara M. Flint. 2003. "Guiding Lights: Intelligence Oversight and Control for the Challenge of Terrorism." *Criminal Justice Ethics* 22, no. 1: 2–58.

BeVier, Lillian R. 1995. "Information and Individuals in the Hands of Government: Some Reflections on Mechanisms for Privacy Protection." *William and Mary Bill of Rights Journal* 4 (Winter): 455–506.

Block, Frederic. 2005. "Civil Liberties during National Emergencies: The Interactions between the Three Branches of Government in Coping with Past and Current Threats to the Nation's Security." *New York University Review of Law and Social Change* 29: 459–524.

Bradley, Alison A. 2002. "Extremism in the Defense of Liberty?: The Foreign Intelligence Surveillance Act and the Significance of the USA PATRIOT Act." *Tulane Law Review* 77, no. 2: 465–93.

Branscomb, Anne Wells. 1994. *Who Owns Information? From Privacy to Public Access*. New York: Basic Books.

Brooks, Rosa Ehrenreich. 2004. "War Everywhere: Rights, National Security Law, and the Law of Armed Conflict in the Age of Terror." *University of Pennsylvania Law Review* 153 (December): 675–761.

Brown, Cynthia, ed. 2003. *Lost Liberties: Ashcroft and the Assault on Personal Freedom*. New York: New Press.

Cassel, Elaine. 2004. *The War on Civil Liberties: How Bush and Ashcroft Have Dismantled the Bill of Rights*. Chicago: Lawrence Hill Books.

Cate, Fred H. 1997. *Privacy in the Information Age*. Brookings.

_____. 2002. *Privacy and Other Civil Liberties in the United States after September 11*. Washington: American Institute of Contemporary German Studies.

Cate, Fred H., and Robert E. Litan. 2002. "Constitutional Issues in Information Privacy." *Michigan Technology and Telecommunications Law Review* 9, no. 1: 35–63.

Chang, Nancy. 2002. *Silencing Political Dissent: How Post–September 11 Anti-Terrorism Measures Threaten Our Civil Liberties.* New York: Seven Stories Press.

Chesney, Robert M. 2003. "Civil Liberties and the Terrorism Prevention Paradigm: The Guilt by Association Critique." *Michigan Law Review* 101, no. 6: 1408–52.

Cohen, David B., and John W. Wells, eds. 2004. *American National Security and Civil Liberties in an Era of Terrorism.* New York: Palgrave Macmillan.

Cole, David. 2003a. *Enemy Aliens: Double Standards and Constitutional Freedoms in the War on Terrorism.* New York: New Press.

____. 2003b. "The New McCarthyism: Repeating History in the War on Terrorism." *Harvard Civil Rights and Civil Liberties Law Review* 38, no. 1: 1–30.

____. 2003c. "Their Liberties, Our Security: Democracy and Double Standards." *International Journal of Legal Information* 31, no. 2: 290–311.

Cole, David, James X. Dempsey, and Carole Goldberg. 2002. *Terrorism and the Constitution: Sacrificing Civil Liberties in the Name of National Security.* New York: New Press.

Connell, Christopher. 2002. *Homeland Defense and Democratic Liberties: An American Balance in Danger?* New York: Carnegie Corporation.

Cooper, Jonathan. 1998. *Liberating Cyberspace: Civil Liberties, Human Rights and the Internet.* London: Pluto Press.

Copeland, Rebecca A. 2004. "War on Terrorism or War on Constitutional Rights? Blurring the Lines of Intelligence Gathering in Post–September 11 America." *Texas Tech Law Review* 35, no. 1: 1–31.

Cothran, Helen. 2004. *National Security: Opposing Viewpoints.* San Diego: Greenhaven Press.

Darmer, M. Katherine B., Robert M. Baird, and Stuart E. Rosenbaum. 2004. *Civil Liberties vs. National Security in a Post-9/11 World.* Amherst, N.Y.: Prometheus Books.

Dash, Samuel. 2004. *The Intruders: Unreasonable Searches and Seizures from King John to John Ashcroft.* Rutgers University Press.

Davis, Darren W., and Brian D. Silver. 2004. "Civil Liberties vs. Security: Public Opinion in the Context of the Terrorist Attacks on America." *American Journal of Political Science* 48, no. 1: 28–46.

Davis, Robert N. 2003. "Striking the Balance: National Security vs. Civil Liberties." *Brooklyn Journal of International Law* 29, no. 1: 175–238.

DeCew, Judith. 1997. *In Pursuit of Privacy: Law, Ethics, and Technology.* Cornell University Press.

Dempsey, James X., and Lara M. Flint. 2004. "Commercial Data and National Security." *George Washington Law Review* 72 (August): 1459–1502.

Dershowitz, Alan M. 2002. *Shouting Fire: Civil Liberties in a Turbulent Age.* Boston: Little, Brown.

Devins, Neal. 2003. "Congress, Civil Liberties, and the War on Terrorism." *William and Mary Bill of Rights Journal* 11, no. 3: 1139–54.

Dinh, Viet. 2002. "Foreword: Freedom and Security after September 11." *Harvard Journal of Law and Public Policy* 25, no. 2: 399, 405–06.

Ditzion, Robert. 2004. "Note. Electronic Surveillance in the Internet Age: The Strange Case of Pen Registers." *American Criminal Law Review* 41, no. 2: 1321–52.

Doherty, Fiona, and Michael McClintock. 2002. *A Year of Loss: Reexamining Civil Liberties Since September 11.* Washington: Lawyers Committee for Human Rights.

Doyle, Charles. 2001."Terrorism: Section by Section Analysis of the USA PATRIOT Act." Report RL31200. Congressional Research Service, Library of Congress.

____. 2002. "The USA PATRIOT Act: A Legal Analysis." Report RL31377. Congressional Research Service.

____. 2004. "USA Patriot Act Sunset: Provisions That Expire on December 31, 2005." Report RL32186. Congressional Research Service.

Dworkin, Ronald. 2002. "The Threat to Patriotism." *New York Review of Books,* February 28, 44–49.

Etzioni, Amitai. 2004. *How Patriotic is the Patriot Act? Freedom Versus Security in the Age of Terrorism.* New York: Routledge.

Etzioni, Amitai, and Jason H. Marsh, eds. 2003. *Rights vs. Public Safety after 9/11: America in the Age of Terrorism.* Lanham, Md.: Rowman and Littlefield.

Evans, Jennifer C. 2002. "Hijacking Civil Liberties: The USA PATRIOT Act of 2001." *Loyola University Chicago Law Journal* 33, no. 4: 933–90.

Federal Trade Commission. 1998. "Privacy Online: A Report to Congress" (www.ftc.gov/reports/privacy3/toc.htm [November 2005]).

____. 2000. "Privacy Online: Fair Information Practices in the Electronic Marketplace—A Report to Congress" (www.ftc.gov/reports/privacy2000/privacy2000.pdf [November 2005]).

Feingold, Russ. 2003. "Introduction." Symposium Issue: Civil Liberties in a Time of Terror. *Wisconsin Law Review* no 2: 255–56.

Fishman, Clifford S. 1987. "Interception of Communications in Exigent Circumstances: The Fourth Amendment, Federal Legislation, and the United States Department of Justice." *Georgia Law Review* 22, no. 1: 1–87.

Freiwald, Susan. 2004. "Online Surveillance: Remembering the Lessons of the Wiretap Act." *Alabama Law Review* 56, no. 9: 46–52.

Froomkin, Michael. 2000. "The Death of Privacy?" *Stanford Law Review* 52, no. 5: 1461–1543.

Gede, Tom, Montgomery N. Kosma, and Arun Chandra. 2001. *White Paper on Anti-Terrorism Legislation: Surveillance and Wiretap Laws. Developing Nec-*

essary and Constitutional Tools for Law Enforcement. Washington: Federalist Society for Law and Public Policy Studies.

Goldstein, Bruce D. 1992. "Confidentiality and Dissemination of Personal Information: An Examination of State Laws Governing Data Protection." *Emory Law Journal* 41 (Fall): 1185–1280.

Grier, Manton M. Jr. 2001. "The Software Formerly Known as Carnivore: When Does E-mail Surveillance Encroach upon a Reasonable Expectation of Privacy?" *South Carolina Law Review* 52, no. 4: 875–94.

Gross, Emanuel. 2004. "The Struggle of a Democracy against Terrorism—Protection of Human Rights: The Right to Privacy versus the National Interest— the Proper Balance." *Cornell International Law Journal* 37, no. 1: 28–93.

Gutterman, Melvin. 1988. "A Formulation of the Value and Means Model of the Fourth Amendment in the Age of Technologically Enhanced Surveillance." *Syracuse Law Review* 39, no. 2: 647–736.

Halperin, Morton H. 1981. *National Security and Civil Liberties: A Benchmark Report.* Washington: Center for National Security Studies.

Hardin, David. 2003. "The Fuss over Two Small Words: The Unconstitutionality of the USA PATIROT Act Amendments to FISA under the Fourth Amendment." *George Washington Law Review* 71, no. 2: 291–345.

Henderson, Nathan C. 2002. "The Patriot Act's Impact on the Government's Ability to Conduct Electronic Surveillance of Ongoing Domestic Communications." *Duke Law Journal* 52 (October): 179–209.

Heymann, Philip B. 2002. "Civil Liberties and Human Rights in the Aftermath of September 11." *Harvard Journal of Law and Public Policy* 25, no. 2: 441–56.

____. 2003. *Terrorism, Freedom, and Security : Winning without War.* MIT Press.

Hoofnagle, Chris Jay. 2004. "Big Brother's Little Helpers: How ChoicePoint and Other Commercial Data Brokers Collect and Package Your Data for Law Enforcement." *North Carolina Journal of International Law and Commercial Regulation* 29 (Summer): 595–637.

Howell, Beryl A. 2004. "USA PATRIOT Act: Seven Weeks in the Making." *George Washington Law Review* 72, no. 6: 1145–1207.

Hufstedler, Shirley M. 1979. "Invisible Searches for Intangible Things: Regulation of Government Information Gathering." *University of Pennsylvania Law Review* 127: 1483, 1510–11.

Jasper, Margaret C. 2003. *Privacy and the Internet: Your Expectations and Rights under the Law.* Dobbs Ferry, N.Y. : Oceana Publications.

Karas, Stan. 2002. "Privacy, Identity, Databases." *American University Law Review* 52, no. 2: 393–445.

Kendal, Aaron. 2001. "Carnivore: Does the Sweeping Sniff Violate the Fourth Amendment?" *Thomas M. Cooley Law Review* 18, no. 2: 183–200.

Kennedy, Charles H., and Peter P. Swire. 2003. "State Wiretaps and Electronic Surveillance after September 11." *Hastings Law Journal* 54, no. 4: 971–986.

Kerr, Donald M. 2000. "Prepared Statement of Donald M. Kerr, Assistant Director, Federal Bureau of Investigation, before the United States Senate, Committee on the Judiciary, September 6, 2000" (commdocs.house.gov/committees/judiciary/hju67343.000/hju67343_0.htm [November 2005]).

Kerr, Orin S. 2001. "The Fourth Amendment in Cyberspace: Can Encryption Create a Reasonable Expectation of Privacy?" *Connecticut Law Review* 33, no. 2: 503–33.

———. 2003. "Internet Surveillance Law after the USA PATRIOT Act: The Big Brother That Isn't." *Northwestern University Law Review* 97 (Winter): 607–73.

———. 2004. "A User's Guide to the Stored Communications Act, and a Legislator's Guide to Amending It." *George Washington Law Review* 72 (August): 1208–43.

Kollar, Justin F. 2004. "USA PATRIOT Act, the Fourth Amendment, and Paranoia: Can They Read this While I'm Typing?" *Journal of High Technology Law* 3, no. 1: 67–93.

Kreimer, Seth F. 2004. "Watching the Watchers: Surveillance, Transparency and Political Freedom in the War on Terror." *University of Pennsylvania Journal of Constitutional Law* 7 (September): 133–81.

Ku, Raymond Shih Ray. 2002. "The Founders' Privacy: The Fourth Amendment and the Power of Technology Surveillance." *Minnesota Law Review* 86, no. 6: 1325–78 .

Lee, Laurie Thomas. 2003. "The USA Patriot Act and Telecommunications: Privacy under Attack." *Rutgers Computer and Technology Law Journal* 29 (June): 371–403.

Leone, Richard C., and Greg Anrig Jr. 2003. *The War on Our Freedoms: Civil Liberties in an Age of Terrorism*. New York: Century Foundation.

Levin, Daniel B., ed. 2003. *U.S. National Security and Civil Liberties*. Oxford: Routledge/Taylor and Francis.

Levine, Burton. 2001. "Terrorism in the Land of the Free: Repression and Incarceration in the Name of Security" *Humanist*. 61, no. 1: 11–13.

Lobel, Jules. 2003. "The War on Terrorism and Civil Liberties." *University of Pittsburgh Law Review* 63, no. 4: 767–90.

Luban, David. 2002. "The War on Terrorism and the End of Human Rights." *Philosophy and Public Policy Quarterly* 22 (Summer): 9–14.

Lynch, Timothy. 2002. *Breaking the Vicious Cycle: Preserving Our Liberties While Fighting Terrorism*. Washington: Cato Institute.

Lytton, Christopher H. 2003. "America's Borders and Civil Liberties in a Post-September 11th World." *Journal of Transnational Law and Policy* 12, no. 2: 197–216.

Mack, Raneta Lawson. 2004. *Equal Justice in the Balance: America's Legal Responses to the Emerging Terrorist Threat*. University of Michigan Press.

Madrinan, Peter G. 2003. "Devil in the Details: Constitutional Problems Inherent in the Internet Surveillance Provisions of the USA Patriot Act of 2001." *University of Pittsburgh Law Review* 64 (Summer): 783–834.

Margulies, Peter. 2004. "The Clear and Present Internet: Terrorism, Cyberspace, and the First Amendment." *UCLA Journal of Law and Technology* (www.lawtechjournal.com/articles/2004/04_041207_margulies.php [December 2005]).

Markle Foundation Task Force on National Security in the Information Age. 2002. *Protecting America's Freedom in the Information Age* (www.markle.org/downloadable_assets/nstf_full.pdf [October 2002]).

———. 2003. *Creating a Trusted Information Network for Homeland Security* (www.markle.org/downloadable_assets/nstf_report2_full_report.pdf [December 2003]).

Mart, Susan Nevelow. 2004. "Protecting the Lady from Toledo: Post–USA Patriot Act Electronic Surveillance at the Library." *Law Library Journal* 96 (Summer): 449–73.

Martin, Kate. 2004. "Intelligence and Civil Liberties." *SAIS Review* 24, no. 1: 7–21.

Mayer, Jeremy D. 2002. "9-11 and the Secret FISA Court: From Watchdog to Lapdog." *Case Western Reserve International Journal of Law* 34, no. 2: 249–52.

McCarthy, Thomas R. 2001. "Don't Fear Carnivore: It Won't Devour Individual Privacy." *Missouri Law Review* 66, no. 4: 827–47.

McLean, Deckle. 1995. *Privacy and Its Invasion*. Westport, Conn.: Praeger.

Mueller, Robert S., III. 2003. "The FBI's New Mission: Preventing Terrorist Attacks While Protecting Civil Liberties." *Stanford Journal of International Law* 39, no. 1: 117-23.

Muller, Eric. 2002. "12/7 and 9/11: War, Liberties, and the Lessons of History." *West Virginia Law Review* 104, no. 3: 571–92.

Mulligan, Deirdre K. 2004. "Reasonable Expectations in Electronic Communications: A Critical Perspective on the Electronic Communications Privacy Act." *George Washington Law Review* 72 (August): 1557–98.

Murray, Nancy. 2002. "Civil Liberties in Times of Crisis: Lessons From History." *Massachusetts Law Review* 87, no. 2: 72–83.

Musch, Donald. 2003. *Civil Liberties and the Foreign Intelligence Surveillance Act*. Dobbs Ferry, N.Y. : Oceana Publications.

Nagan, Winston P., and Craig Hammer. 2004. "The New Bush National Security Doctrine and the Rule of Law." *Berkeley Journal of International Law* 22, no. 3: 375–438.

National Commission on Terrorist Attacks. 2004. *The 9/11 Commission Report: Final Report of the National Commission on Terrorist Attacks upon the United States*. New York: W.W. Norton.

Osher, Steven A. 2002. "Privacy, Computers and the Patriot Act: The Fourth Amendment Isn't Dead, but No One Will Insure It." *Florida Law Review* 54, no. 3: 521–42.

Posner, Richard. 1978. "The Right to Privacy." *Georgia Law Review* 12, no. 3: 393–422.

___. 2003. *Law, Pragmatism and Democracy*. Harvard University Press.

Rackow, Sharon H. 2002. "How the USA PATRIOT Act Will Permit Governmental Infringement upon the Privacy of Americans in the Name of 'Intelligence Investigations.'" *University of Pennsylvania Law Review* 150, no. 5: 1651–96.

Raul, Alan Charles. 2002. *Privacy and the Digital State: Balancing Public Information and Personal Privacy*. Washington: Progress and Freedom Foundation.

Richards, Robert, and Clay Calvert. 2003. "Nadine Strossen and Freedom of Expression: A Dialogue with the ACLU's Top Card-Carrying Member." *George Mason University Civil Rights Law Journal* 13, no. 2: 185–241.

Robinson, Gerald H. 2000. "We're Listening—Electronic Eavesdropping, FISA, and the Secret Court." *Willamette Law Review* 36, no. 1: 51–81.

Rodriguez, Alejandra. 2003. "Is the War on Terrorism Compromising Civil Liberties? A Discussion of Hamdi and Padilla." *California Western Law Review* 39, no. 2: 389–94.

Rosen, Jeffrey. 2001a. *The Naked Crowd: Reclaiming Security and Freedom in an Anxious Age*. New York: Random House.

____. 2001b. *The Unwanted Gaze: The Destruction of Privacy in America*. New York: Vintage Books.

Rosenzweig, Paul. 2003a. "Balancing Liberty and Security." Commentary. Washington: Heritage Foundation (May 14).

____. 2003b. "Proposals for Implementing the Terrorism Information Awareness System." Legal Memorandum 8. Heritage Foundation (August 7).

____. 2004. "Liberty and the Response to Terrorism." *Duquesne University Law Review* 42 (Summer): 663–723.

Rosenzweig, Paul, and Ha Nguyen. 2003. "CAPPS II Should Be Tested and Deployed." Backgrounder 1683. Heritage Foundation (August 28).

Rotenberg, Marc. 2001 "Fair Information Practices and the Architecture of Privacy (What Larry Doesn't Get)." *Stanford Technology Law Review* (stlr.stanford.edu/STLR/Articles/01_STLR_1/article_pdf.pdf [December 2005]).

____. 2002. "Privacy and Secrecy after September 11." *Minnesota Law Review* 86, no. 6: 1115–36.

Rubin, Paul H., and Thomas M. Leonard. 2002. *Privacy and the Commercial Use of Personal Information*. Washington: Progress and Freedom Foundation.

Safire, William. 2002. "You are a Suspect." *New York Times*, November 14, p. A35.

Schulhofer, Stephen J. 2002. *The Enemy Within: Intelligence Gathering, Law Enforcement, and Civil Liberties in the Wake of September 11*. New York: Century Foundation.

Schultz, Christian David Hamel. 2001. "Unrestricted Federal Agent: 'Carnivore' and the Need to Revise the Pen Register Statute." *Notre Dame Law Review* 76, no. 4: 1215–59.

Schwartz, Paul M. 2004. "Evaluating Telecommunications Surveillance in Germany: The Lessons of the Max Planck Institute's Study." *George Washington Law Review* 72 (August): 1244–63.

Sidel, Mark. 2004. *More Secure, Less Free? Antiterrorism Policy and Civil Liberties after September 11*. University of Michigan Press.

Singleton, Solveig. 2004. "Privacy and Twenty-First Century Law Enforcement: Accountability for New Techniques." *Ohio Northern University Law Review* 30, no. 3: 417–50.

Slobogin, Christopher. 1997. "Technologically Assisted Physical Surveillance: The American Bar Association's Tentative Draft Standards." *Harvard Journal of Law and Technology* 10, no. 3: 383–464.

____. 2002. "Peeping Techno-Toms and the Fourth Amendment: Seeing through Kyllo's Rules Governing Technological Surveillance." *Minnesota Law Review* 86, no. 6: 1393–1438.

Smith, Marcia S. 2003. "Internet Privacy: Overview and Pending Legislation." Report RL31408. Congressional Research Service.

Smith, Marcia S., and others. 2002. "The Internet and the USA PATRIOT Act: Potential Implications for Electronic Privacy, Security, Commerce, and Government." Report RL31289. Congressional Research Service.

Solove, Daniel J. 2002a. "Conceptualizing Privacy." *California Law Review* 90, no. 4: 1097–1156.

____. 2002b. "Digital Dossiers and the Dissipation of Fourth Amendment Privacy." *Southern California Law Review* 75, no. 5: 1083–1168.

____. 2004a. "Restructuring Electronic Surveillance Law." *George Washington Law Review* 72 (August): 1264–1305.

____. 2004b. *The Digital Person: Technology and Privacy in the Information Age*. NYU Press.

Steinberg, David. 1990. "Making Sense of Sense-Enhanced Searches." *Minnesota Law Review* 74, no. 3: 563–629.

Steinberg, James, Mary Graham, and Andrew Eggers. 2003. "Building Intelligence to Fight Terrorism." Policy Brief 125 (Brookings, September).

Stevens, Gina Marie. 2003. "Privacy: Total Information Awareness Programs and Related Information Access, Collection and Protection Laws." Report RL31730. Congressional Research Service.

Stone, Geoffrey R. 2003. "Civil Liberties in Wartime." *Journal of Supreme Court History* 28, no. 3: 215–51.

____. 2004a. *Perilous Times: Free Speech in Wartime*. New York: W.W. Norton.

____. 2004b. "War Fever." *University of Missouri Law Review* 69 (Fall): 1131–55.

Strickland, Lee S. 2002. "Information and the War against Terrorism. Part IV. Civil Liberties versus Security in the Age of Terrorism." *Bulletin of the American Society for Information Science* 28, no. 4: 9–13.

Strickland, Lee S., David A. Baldwin, and Marlene Justsen. 2004. "Domestic Security Surveillance and Civil Liberties." In *Annual Review of Information*

Science and Technology 2005, vol. 39, edited by Blaise Cronin, chap. 11. Medford, N.Y.: Information Today.

Stuntz, William J. 1995. "Privacy's Problem and the Law of Criminal Procedure." *Michigan Law Review* 93 (March): 1016–78.

Swire, Peter. 2001. "Administration Wiretap Proposal Hits Right Issues but Goes Too Far." Analysis Paper 3 (Brookings, October 3).

____. 2004. "The System of Foreign Intelligence Surveillance Law." *George Washington Law Review* 72 (August): 1306–71.

Taipale, Kim. A. 2003. "Data Mining and Domestic Security: Connecting the Dots to Make Sense of Data." *Columbia Science and Technology Law Review* (www.stlr.org/html/volume5/taipaleintro.html [December 2005]).

____. 2004–05. "Technology, Security and Privacy: The Fear of Frankenstein, The Mythology of Privacy and the Lessons of King Ludd." *Yale Journal of Law and Technology* 7 (Fall): 123–221.

Taylor, Stuart Jr. 2003. "Rights, Liberties, and Security—Recalibrating the Balance after September 11." *Brookings Review* 21, no. 1: 25–31.

Technology and Privacy Advisory Committee. 2004. *Safeguarding Privacy in the Fight Against Terrorism*. U.S. Department of Defense (March).

Terbrusch, Richard P. 2004. "Gathering Foreign Intelligence in Cyberspace: Does the United States Need Another Secret Court?" *Quinnipiac Law Review* 23, no. 3: 989–1025.

Thompson, Larry D. 2003. "Intelligence Collection and Information Sharing within the United States" (www.brookings.edu/views/testimony/thompson/20031208.htm [November 2005]).

Treverton, Gregory. 2001. *Reshaping National Intelligence for an Age of Information*. Cambridge University Press.

Tushnet, Mark. 2003. "Defending Korematsu? Reflections on Civil Liberties in Wartime." *Wisconsin Law Review* 2, no. 2: 273–307.

U.S. Department of Defense. 2003. *Report to Congress Regarding the Terrorism Information Awareness Program*. May 20, 2003.

U.S. Department of Justice. Inspector General. 2003. "The September 11 Detainees: A Review of the Treatment of Aliens Held on Immigration Charges in Connection with the Investigation of the September 11 Attacks" (justice.gov/oig/special/0306/index.htm [November 2005]).

U.S. General Accounting Office. 2000. *Federal Agencies' Fair Information Practices*. GAO/AIMD-00-296R.

____. 2003. *Information Technology: Terrorist Watch Lists Should Be Consolidated to Promote Better Integration and Sharing*. GAO-03-322.

U.S. Senate Committee on the Judiciary. 2003. "Interim Report on FBI Oversight: FISA Implementation Failures" (www.fas.org/irp/congress/2003_rpt/ fisa.html [November 2005]).

U.S. Senate Select Committee on Intelligence and U.S. House Permanent Select Committee on Intelligence. 2002. *Joint Inquiry into Intelligence Community Activities before and after the Terrorist Attacks of September 11, 2001*. S.

Rept. 107–351, H. Rept. 107-792. 107 Cong, 2 sess. Government Printing Office.

Van Alstyne, William W. 2003. "Civil Rights and Civil Liberties: Whose Rule of Law?" *William and Mary Bill of Rights Journal* 11, no. 2: 623–53.

Verton, Dan. 2003. *Black Ice: The Invisible Threat of Cyber-terrorism*. New York: McGraw-Hill/Osborne.

Wade, Lindsey. 2003. "Terrorism and the Internet: Resistance in the Information Age." *Knowledge, Technology, and Policy* 16, no. 1: 104–27.

Watanabe, Nathan. 2003. "Internment, Civil Liberties, and a National Crisis." *Southern California Interdisciplinary Law Journal* 13, no. 1: 167–93.

Weich, Ronald. 2002. "Insatiable Appetite: The Government's Demand for New and Unnecessary Powers after September 11" (www.aclu.org/safefree/resources/17042pub20021015.html#attach [November 2005]).

Wells, Christina E. 2004. "Information Control in Times of Crisis: The Tools of Repression." *Ohio Northern Law Review* 30, no. 3: 451–94.

Westin, Alan F. 1967. *Privacy and Freedom*. New York: Atheneum.

Whitaker, Reg. 1999. *The End of Privacy: How Total Surveillance Is Becoming a Reality*. New York: New Press.

Whitehead, John W., and Steven H. Aden. 2002. "Forfeiting Enduring Freedom for Homeland Security: A Constitutional Analysis of the USA Patriot Act and the Justice Department's Anti-Terrorism Initiatives." *American University Law Review* 51, no. 6: 1081–1133.

Contributors

ZOË BAIRD is president of the Markle Foundation, a private philanthropy that focuses on using information and communications technologies to address critical public needs, and is cochair of the Markle Task Force on National Security. Baird previously served as a senior vice president and general counsel of Aetna, a senior visiting scholar at Yale Law School, a counselor and staff executive at General Electric, and a partner in the law firm of O'Melveny and Myers. She also served in the administrations of Presidents Carter and Clinton and was a member of the Technology and Privacy Advisory Committee, which advised the Defense Department on the use of technology to fight terrorism.

JAMES BARKSDALE is a partner and cofounder of the Barksdale Group and is cochair of the Markle Task Force on National Security. Before founding the Barksdale Group, he served as president and CEO of Netscape Communications Corporation. Before Netscape he worked at AT&T Wireless Services as the company's CEO, and before that he served as executive vice president and COO for Federal Express Corporation. Barksdale is presently a member of the President's Foreign Intelligence Advisory Board.

BRUCE BERKOWITZ is a research fellow at the Hoover Institution at Stanford University and is author of many books and articles on national security affairs, including his most recent, *The New Face of War: How War Will Be Fought in the 21st Century*.

JERRY BERMAN is president of both the Center for Democracy and Technology (CDT), an Internet policy and civil liberties organization, and the Internet Education Foundation. He also chairs the Congressional Internet Caucus Advisory Committee (which includes 150 organizations). Before founding CDT, Berman was director of the Electronic Frontier Foundation and chief legislative counsel at the ACLU. He helped to craft the Foreign Intelligence Surveillance Act of 1978 and the Electronic Communications Act of 1986, among other laws.

RUSS FEINGOLD was elected to the U.S. Senate from Wisconsin in 1992 and reelected in 1998 and 2004. Senator Feingold serves on the Senate Judiciary Committee and the Foreign Relations Committee and is ranking member of the Subcommittee on the Constitution.

BERYL A. HOWELL is the managing director and general counsel of the Washington, D.C., office of Stroz Friedberg, a professional consulting and technical services firm specializing in computer forensics and cybersecurity investigations. Howell is a former general counsel of the Senate Committee on the Judiciary, where she worked for Senator Patrick Leahy (D-Vt.).

JON KYL was elected to the U.S. Senate from Arizona in 1994 and reelected in 2000, after having served four terms in the U.S. House of Representatives. Senator Kyl serves on the Senate Judiciary Committee and the Senate Finance Committee and is chairman of the Subcommittee on Terrorism, Technology, and Homeland Security.

GILMAN LOUIE is CEO of In-Q-Tel, the private technology venture group funded by the Central Intelligence Agency. Before helping found In-Q-Tel, Louie served as Hasbro Interactive's chief creative officer and as general manager of the Games.com group, for which he helped pioneer and market such interactive entertainment products as *Falcon*, the F-16 flight simulator, and *Tetris*, which he discovered in the Soviet Union. Before joining Hasbro, he served as chief executive of the Nexa Corporation, Sphere, Spectrum HoloByte, and MicroProse. He has also served on the board of directors of Total Entertainment Network, Direct Language, and FASA Interactive.

CLAYTON NORTHOUSE is an information policy analyst at OMB Watch and program director at the Computer Ethics Institute.

JAMES B. STEINBERG is dean of the LBJ School of Public Affairs, University of Texas at Austin, and served as vice president and director of the Foreign Policy Studies program at the Brookings Institution from 2001 through 2005. Steinberg was deputy national security adviser to President Clinton from December 1996 to July 2000.

LARRY THOMPSON is senior vice president and general counsel of Pepsico. He served as deputy attorney general in the Bush administration from 2001 to 2003. In addition, he led the National Security Coordination Council, which works to ensure seamless coordination of all functions of the Justice Department relating to national security, particularly its efforts to combat terrorism. Thompson was also a senior fellow at the Brookings Institution, was partner in the law firm King and Spalding, and was the U.S. attorney for the Northern District of Georgia.

GAYLE VON ECKARTSBERG is vice president for Strategy and Communications at In-Q-Tel and has served at the CIA, Department of State, and the Office of the Secretary of Defense, International Security Affairs.

ALAN F. WESTIN is professor emeritus of Public Law and Government at Columbia University and president of the nonprofit Center for Social and Legal Research (CSLR). An expert on privacy for over forty years (author of *Privacy and Freedom* and *Databanks in a Free Society*), he has been academic advisor to over fifty national surveys on privacy issues, four of which have explored the security-versus-liberty tension since September 11. Currently, he is director of CSLR's Center for Strategic Privacy Studies, conducting research on post–September 11 antiterrorist issues involving use of government powers of investigation, monitoring, and surveillance.

Index